Jingqiao Zhang and Arthur C. Sanderson

Adaptive Differential Evolution

Adaptation, Learning, and Optimization Volume 1

Further volumes of this series can be found on our homepage: springer.com

Vol. 1. Jingqiao Zhang and Arthur C. Sanderson
Adaptive Differential Evolution, 2009
ISBN 978-3-642-01526-7

Jingqiao Zhang and Arthur C. Sanderson

Adaptive Differential Evolution

A Robust Approach to Multimodal Problem Optimization

Springer

Dr. Jingqiao Zhang
Electrical, Computer and System Engineering Department
Rensselaer Polytechnic Institute
Troy, NY 12180
USA
E-mail: zhangj14@rpi.edu

Dr. Arthur C. Sanderson
Electrical, Computer and System Engineering Department
Rensselaer Polytechnic Institute
Troy, NY 12180
USA
E-mail: sandea@rpi.edu

ISBN 978-3-642-01526-7 e-ISBN 978-3-642-01527-4

DOI 10.1007/978-3-642-01527-4

Adaptation, Learning, and Optimization ISSN 1867-4534

Library of Congress Control Number: Applied for

Typeset : Scientific Publishing Services Pvt. Ltd., Chennai, India.

Printed in acid-free paper

9 8 7 6 5 4 3 2 1

springer.com

Foreword

I first met Jingqiao when he had just commenced his PhD research in evolutionary algorithms with Arthur Sanderson at Rensselaer. Jingqiao's goals then were the investigation and development of a novel class of self-adaptive differential evolution algorithms, later called JADE. I had remarked to Jingqiao then that Arthur always appreciated strong theoretical foundations in his research, so Jingqiao's prior mathematically rigorous work in communications systems would be very useful experience. Later in 2007, when Jingqiao had completed most of the theoretical and initial experimental work on JADE, I invited him to spend a year at GE Global Research where he applied his developments to several interesting and important real-world problems.

Most evolutionary algorithm conferences usually have their share of innovative algorithm oriented papers which seek to best the state of the art algorithms. The best algorithms of a time-frame create a foundation for a new generation of innovative algorithms, and so on, fostering a meta-evolutionary search for superior evolutionary algorithms. In the past two decades, during which interest and research in evolutionary algorithms have grown worldwide by leaps and bounds, engaging the curiosity of researchers and practitioners from many diverse science and technology communities, developing stand-out algorithms is getting progressively harder.

JADE stands out as a unique algorithm for these principal reasons: it works very well, it has an elegant theoretical foundation, it has been recognized as the winner in the 2008 IEEE World Congress on Computational Intelligence competition, and it is very user-friendly relieving the user of the task of algorithm parameter setting. As most decision and control problems may be cast in the semantics of an optimization or search problem, powerful new methods that can efficiently handle nonlinear, discontinuous, and high-dimensional spaces have a very significant real-world impact.

The No Free Lunch Theorem for optimization states that for any optimization algorithm, an elevated performance over one class of problems is offset by degraded performance over another class. Thus, the performance average of a given optimization algorithm over the entire class of potential problems is constant. If an algorithm performs better than random search for some problems, it will perform worse than random search for other problems, maintaining a constant performance average. These results suggest that any one set of potentially optimal algorithms can be considered so only for a limited subset of problems and their use cannot result in consistently superior performance over the entire space of optimization problems, as was previously generally expected. However, an algorithm can outperform another if it is specialized to the structure of the specific problem class under consideration. An alternative approach, as developed in JADE, is to embed self-adaptation in the search strategy that can automatically adapt control parameters in a manner responsive to a wide range of search space characteristics and at different stages of the evolutionary search. This need for adaptively evolving algorithm parameters as the evolutionary search proceeds is highlighted by the theoretical results underpinning JADE. Further, the self-adaptation characteristic of JADE enhances its usability as the user does not need to have significant knowledge of the interaction between control parameters and algorithm behavior.

The book's authors present an excellent balance of rigor, clarity, and real-world focus. This book, for the technical reader, is an exciting document covering a powerful class of state of the art self-adaptive differential evolution algorithms, JADE, that has demonstrated its mettle not only on several standard experimental test spaces, but also importantly on several complex real-world optimization problems spanning combinatorial auctions, financial credit decision-making, and flight route planning in air-traffic control systems.

Dr. Raj Subbu
Senior Computer Scientist
General Electric Global Research
Niskayuna, NY 12309, USA

Acknowledgements

The authors would like to thank Springer publishers and series editor Dr. Meng-Hiot Lim for selecting this work as a part of the series on Evolutionary Learning and Optimization.

This book is based on the doctoral research by the first author during the years 2005–2008 under the guidance of his advisor Dr. Arthur C. Sanderson, the second author, at Rensselaer Polytechnic Institute. The authors would like to acknowledge support of this research by grant number IIS-0329837 from the National Science Foundation. This work was also supported in part by the Center for Automation Technologies and Systems (CATS) under a block grant from the New York State Office of Science, Technology, and Academic Research (NYSTAR). The authors thank Dr. Murat Arcak, Dr. Piero P. Bonissone, Dr. Alan A. Desrochers, Dr. Volkan Isler and Dr. Raj Subbu, for serving on the doctoral dissertation committee and providing valuable feedback and insightful suggestions to improve this work. The authors also thank the researchers at General Electric Global Research for their professional advice and helpful discussions during the first author's internship. There are also many other members of the Center for Automation Technologies and Systems and the Department of Electrical, Computer and Systems Engineering at Rensselaer Polytechnic Institute who contributed their advice and insight to this work.

Finally, on a personal note, the first author would like to express his sincerest thank and gratitude to his advisor Dr. Arthur C. Sanderson and his internship supervisor Dr. Raj Subbu for their continuous guidance, inspiration and encouragement, without which this work would not have been possible.

Contents

Acronyms

AEC	Adaptive Evolution Control
CA	Combinatorial Auction
CC	Cooperative Coevolution
DE	Differential Evolution
EA	Evolutionary Algorithms
EP	Evolutionary Programming
ES	Evolution Strategy
ETPM	Empirical Transition Probability Matrix
FESS	Function Evaluations in Successful Runs
GA	Genetic Algorithm
GP	Genetic Programming
MOEA	Multi-objective Evolutionary Algorithm
PSO	Particle Swarm Optimization
RBF	Radial Basis Function
SNR	Signal-to-noise ratio
SR	Success Rate
SP	Success Performance
TPM	Transition Probability Matrix

Chapter 1
Introduction

Optimization problems are ubiquitous in academic research and real-world applications such as in engineering, finance, and scientific areas. What coefficients of a neural network minimize classification errors? What combination of bids maximizes the outcome in an auction? What variable- and check-node distributions optimize a low-density parity-check code design? In general, optimization problems arise wherever such resources as space, time and cost are limited. With no doubt, researchers and practitioners need an efficient and robust optimization approach to solve problems of different characteristics that are fundamental to their daily work.

It is expected that solving a complex optimization problem itself should not be very difficult; e.g., an engineer with expert knowledge of channel coding does not have to be an expert in optimization theory just to improve his/her code designs. In addition, an optimization algorithm should be able to reliably converge to the true optimum for a variety of different problems. Furthermore, the computing resources spent on searching for a solution should not be excessive. Thus, a useful optimization method should be easy to use, reliable and efficient to achieve satisfactory solutions.

Differential Evolution (DE) is such an optimization approach that addresses these requirements. Since its invention by Storn and Price in 1995 [1], [2], DE has been shown to be a simple yet efficient optimization approach in solving a variety of benchmark problems as well as many real-world applications. Differential evolution, together with evolution strategies (ES) developed by Rechenberg [3] and Schwefel [4], genetic algorithms (GA) by Holland [5] and Goldberg [6], and evolutionary programming (EP) by Fogel [7] etc., can be categorized into a class of population-based, derivative-free methods known as evolutionary algorithms (EAs). All these approaches mimic Darwinian evolution and evolve a population of individuals from one generation to another by analogous evolutionary operations such as mutation, crossover and selection.

Like nearly all EAs, DE is a population-based optimizer that starts the optimization process by sampling the search space at multiple, randomly chosen initial points (i.e., a population of individual vectors). Similar to ES, DE is in nature a derivative-free continuous function optimizer, as it encodes parameters as floating-point numbers

J. Zhang and A.C. Sanderson: Adaptive Differential Evolution, ALO 1, pp. 1–4.
springerlink.com

and manipulates them with simple arithmetic operations such as addition, subtraction, and multiplication. Like other evolutionary algorithms DE generates new points that are the perturbations/mutations of existing points; the perturbations, however, come neither from samples of a predefined probability distribution like those in ES nor from centroid-based difference vectors as used in some nonlinear optimization methods such as the Nelder-Mead method [8]. Instead, DE mutates a (parent) vector in the population with a scaled difference of other randomly selected individual vectors. The resultant mutation vector is crossed over with the corresponding parent vector to generate a trial or offspring vector. Then, in a one-to-one selection process of each pair of offspring and parent vectors, the one with a better fitness value survives and enters the next generation. This procedure repeats for each parent vector and the survivors of all parent-offspring pairs become the parents of a new generation in the evolutionary search cycle. The evolutionary search stops when the algorithm converges to the true optimum or a certain termination criterion such as the number of generations is reached.

1.1 Research Motivation

The basic scheme of differential evolution works generally well for many practical problems when the control parameters involved in mutation and crossover are set appropriately There are some simple guidelines of parameter settings in the literature [1], [9], [10], which however are mainly based on experimental studies. Their applicability to other problems is usually not satisfactory, as it has been shown that the control parameters of DE are quite dependent on problem characteristics; i.e., there is no single parameter setting that is suitable for various problems or works consistently well at different evolution stages of a single problem. For example, experimental studies show that different parameter values should be used for a spherical function and an ellipsoid function, although the latter is a simple linear transformation of the former. Also, in the optimization of a spherical function, our theoretical analyses indicate that the value of a control parameter should be adaptively changed, as the function landscape demonstrates different characteristics during the process of approaching the optimum from remote starting points.

Theoretical analyses have been conducted for evolutionary strategies and genetic algorithms along with their algorithm development for several decades [6], [11], [12]. In contrast, only a few theoretical analyses [2], [13], [14], [15] have been conducted concerning the stochastic properties and convergence behavior of differential evolution. In addition, most of them focus only on mutation and crossover while omitting the selection operation of DE. As it is clear that mutation and crossover are both problem-independent operations, the one-to-one selection operation is immediately related to the function values and landscape characteristics of optimization problems. It is thus necessary to consider the selection in a theoretical analysis to understand the problem dependence of the control parameters.

In view of these considerations, some adaptive or self-adaptive schemes [16], [17], [18], [19], [20], [21], [22], [23], [24], [25] have recently been proposed to automatically update the control parameters of differential evolution according to the feedback from the evolutionary search, which in turn indicates the suitability of the current parameter setting to the landscape characteristics of optimization problems. They avoid users' manual parameter tuning and interaction during the evolutionary search and therefore improve the robustness of the algorithm. As the robustness is considered as a priority, these parameter adaptation schemes usually take the most reliable DE strategy as the basis. So far, the direction information to the optimum, such as the best-solution information or history information that is shown to be beneficial to some greedy DE and other evolutionary algorithms, have not been fully utilized to improve the convergence performance of the algorithm. It is expected that the incorporation of parameter adaptation with the greedy DE strategy is capable of achieving a good balance between the convergence rate and reliability of the algorithm and thus producing better performance.

1.2 Research Contribution

This monograph addresses the effect of control parameters of differential evolution and focuses on theoretical and algorithmic development of parameter adaptive differential evolution. The major contributions are summarized below.

First, a theoretical analysis has been conducted to investigate the evolutionary stochastic properties of a differential evolution algorithm's convergence behavior for a spherical function model [26]. A Gaussian approximate model of differential evolution is introduced to facilitate mathematical derivation of the stochastic properties of mutation and selection and help the understanding of the effect of mutation factor on the evolutionary dynamics. Theoretical results, as verified by experimental simulations, highlight the necessity to adaptively evolve control parameters as evolutionary search proceeds.

Second, a new parameter adaptive differential evolution algorithm, JADE, is introduced to automatically adapt control parameters responding to a range of function characteristics at different stages of evolutionary search [27], [28]. To improve convergence performance, JADE incorporates a greedy mutation strategy that utilizes the direction information provided by both high-quality solutions in the current generation and inferior solutions previously explored in the evolutionary search. Extensive experiments have been conducted to show their mutual benefits for a good balance between the convergence speed and reliability. In addition, the parameter adaptation avoids the need for users' prior knowledge of the interaction between the control parameters and the convergence behavior, thus making the algorithm easily usable.

Third, extensions of the basic JADE algorithm are considered from several perspectives. Some optimization problems have a demanding limitation on the number of original function evaluations that are expensive in terms of time, cost and/or other limited resources. JADE has been extended to address these problems by incorporating computationally inexpensive surrogate models and adaptively controlling the level of incorporation according to the model accuracy [29]. Radial basis function networks are used to create these models for the sake of a good balance between computational complexity and model accuracy. Experimental results have shown the efficiency of adaptive incorporation of surrogate models in avoiding potential false or premature convergence while significantly reducing the number of original expensive function evaluations.

Fourth, many real-world applications involve multi-objective optimization, because they are linked to competing or conflicting outcomes such as profit, risk and/or many other performance related criteria. JADE is extended to multi-objective optimization to search for a set of optimal tradeoff solutions [30]. Simulation results show that JADE achieves fast convergence rate by utilizing the direction information provided by previously explored inferior solutions, even when most or all solutions in the current population become non-dominated and thus the high-quality solutions among them do not necessarily indicate the direction of the optimum.

Fifth, the performance of JADE has been evaluated for a variety of benchmark functions of different characteristics such as nonlinearity, non-convexity, multimodality and high-dimensionality. The performance of JADE is compared with both classic and state-of-the-art algorithms such as the classic DE, other adaptive DE algorithms (SaDE [21], jDE [22], SaNSDE [25]), the canonical PSO [31], Adaptive LEP and Best Levy [32], [33] for single-objective optimization, and MODE [15], SPEA [58], GDE3 [34] and NSGA-II [35] for multi-objective optimization. Comprehensive experimental studies have shown that JADE achieves consistently better or competitive convergence performance than other algorithms and has better scalability for high dimensional functions [36]. JADE usually leads to the fastest convergence rate while maintaining the probability of successful convergence at a level very close to the most robust competitive algorithm.

The adaptive differential evolution algorithm JADE has been applied in different real-world applications where domain knowledge can be utilized to improve optimization performance. First, it is applied to a winner determination problem in combinatorial auctions [37], which can be generalized to resource allocation problems in sensor management, supply chain management, etc. Second, JADE solves a data-smoothing problem in credit decision making by searching for the transition probability matrix that satisfies different desired properties and optimally matches the empirical data of credit rating transition [38]. Furthermore, JADE is applied to conduct automatic flight planning in air traffic control systems. It is used to assign an appropriate route to each flight to minimize both system level congestion and total flight distance. In these applications, JADE has produced better results than other state-of-the-art evolutionary algorithms and heuristic approaches in terms of the convergence rate and the quality of the final solutions.

Chapter 2
Related Work and Background

This chapter introduces background information on several topics. First, the basic concepts of evolutionary algorithms are overviewed in Sect. 2.1. The procedure of classic differential evolution is then described in Sect. 2.2 to serve as a basis for theoretical analysis and algorithm development in later chapters. Different parameter control mechanisms are summarized in Sect. 2.3. Multi-objective optimization is introduced in Sect. 2.4 as an application domain of parameter adaptive differential evolution. Finally, the no-free lunch theory and domain knowledge utilization are briefly discussed in Sect. 2.5.

2.1 Evolutionary Algorithms

As its names suggests, an evolutionary algorithm is a special computation technique that draws inspiration from the principle of natural evolution and survival of the fittest that are described by the Darwinian Theory [39]. The Darwinian Theory explains the principle of natural selection, which favors the survival of species that are matched with their environmental conditions and explains the evolution of species as an outcome of stochastic variations and natural selection [11]. The major applications of evolutionary algorithms are in optimization, although they have also been used to conduct data mining, generate learning systems, and build experimental frameworks to validate theories about biological evolution and natural selection, etc. EAs differ from traditional optimization techniques in that they usually evolve a population of solutions or individual points in the search space of decision variables, instead of starting from a single point. At each iteration, an evolutionary algorithm generates new (offspring) solutions by mutating and/or recombining current (parent) solutions and then conducts a competitive selection to weeds out poor solutions. In comparison with traditional optimization techniques, such as calculus-based nonlinear programming methods in [40], evolutionary algorithms are usually more robust and achieve a better balance between the exploration and exploitation in the search space when optimizing many real-world problems. They are efficient to solve problems that are characterized by chaos, chance, temporality, and nonlinear interactivity which tend to be intractable to traditional methods [41].

J. Zhang and A.C. Sanderson: Adaptive Differential Evolution, ALO 1, pp. 5–13.
springerlink.com

Different main streams of evolutionary algorithms have evolved over the past forty years. The majority of the current implementations descend from three strongly related but independently developed branches: evolution strategies [3], [4], genetic algorithms [5], [6], and evolutionary programming [7]. These approaches are closely related to each other in terms of their underlying principles, while their exact operations and usually applied problem domains differ from one approach to another. For example, genetic algorithms usually strongly emphasize recombination. They are better suited for discrete optimization because the decision variables are originally encoded as bit strings and are modified by logical operators. Evolution strategies and evolutionary programming, however, concentrate on mutation, although evolution strategies may also incorporate crossover or recombination as an operator. ES is a continuous function optimizer in nature because it encodes parameters as floating-point numbers and manipulates them by arithmetic operations.

Although differences exist among these evolutionary algorithms, they all rely on the concept of a population of individuals or solutions, which undergo such probabilistic operations as mutation, crossover and selection to evolve toward solutions of better fitness in the search space of decision variables. The mutation introduces new information into the population by randomly generating variations to existing individuals. The crossover or recombination typically performs an information exchange between different individuals in the current population. The selection imposes a driving force towards the optimum by preferring individuals of better fitness. The fitness value may reflect the objective function value and/or the level of constraint satisfaction. These operations compose a loop (named a generation), and evolutionary algorithms usually execute a number of generations until the obtained best-so-far solution is satisfactory or other termination criterion is fulfilled.

2.1.1 Evolution Strategies

Evolution strategies (ESs) were developed by Rechenberg [3] and Schwefel [4] to tackle optimization problems in a continuous search space. In the simplest form, a $(1+1)$-ES works on a single individual and employs a mutation operator to create a new offspring individual at each generation. The better of the parent and the offspring survives while the other is discarded. In a population based ES, such as $(\mu+\lambda)$-ES, μ parents create λ offspring by the recombination and mutation. The best μ survivors are chosen from the union of the parents and offspring. This guarantees a monotone improvement on the quality of solutions during the evolutionary search. In another variant of ES, (μ,λ)-ES, $\lambda > \mu \geq 1$, μ parents create λ offspring in a way similar to that in $(\mu+\lambda)$-ES. However, the best μ offspring replace the parents deterministically; thus it is possible that the best individual at generation $g+1$ is worse than that at generation g. This possible acceptance of worse solutions may be useful to avoid false convergence to a local optimum. Both (μ,λ)-ES and $(\mu+\lambda)$-ES are currently the commonly used evolution strategies. In both schemes, each individual in the population is a concatenation of the decision variables and the control parameters (such as mutation rate) of the algorithm. Thus, the control parameters undergo evolutionary operations, such as recombination and mutation, together with the decision variables.

2.1.2 Evolutionary Programming

Evolutionary programming was first developed by Fogel [7] to use simulated evolution as a learning process to generate artificial intelligence. The original EP was designed to operate on finite state machines, and some of its original variants are quite similar to the genetic programming. However, current EP has evolved towards continuous problem optimization. In a continuous domain [42], EP works on a population of $\mu > 1$ parent individuals and generates μ offspring by its main variation operator of mutation. The μ best solutions in the union of the parents and offspring are selected to constitute the parent at the next generation. Thus, EP can be viewed as a special case of a $(\mu + \mu)$-ES without recombination. Current EP methods are very similar to evolution strategies but with some differences in the selection operation.

2.1.3 Genetic Algorithms

Genetic algorithms were developed by Holland in 1970s and significantly extended by De Jong [43] and Goldberg [6]. Traditionally, individuals are represented as binary strings and offspring are created primarily through the recombination of pairs of parent individuals, which are randomly selected with probabilities related to their fitness values. The mutation operation is usually conducted with a low probability. Early popularity of a binary representation is due to Holland's Schema Theory, which tries to analyze GAs in terms of their expected schema sampling behavior [5], [6]. The binary string representation distinguishes GAs from other EA methods and is very convenient for optimization problems in combinatorial area (e.g. the traveling salesman problem). However, in other problem domains, it may require rather complex encoding and decoding functions to map non-binary solutions to binary strings and vice versa. In this case, it is argued that a binary representation has serious disadvantages because the mapping may further complicate an already nontrivial objective function by introducing an additional multimodality [44]. Other representation methods that are more apt to problem domains have received considerable attention [45].

2.1.4 Genetic Programming

Genetic programming [46] is a special evolutionary algorithm where the individuals in the population are computer programs (e.g., mathematical expression, LISP S-expression, parse tree.) It applies the approach of genetic algorithms, while in his seminal work [46], Koza posits GP as a generalization of genetic algorithms rather than a specialization. Computer programs, possibly written in any programming languages, can be considered as a sequence of applications of functions (e.g., arithmetic operations, standard programming operations, logical functions, or domain-specific functions) to values (variables or constants). The programs are usually expressed in tree structures rather than lines of codes. In this way, the variables and constants are represented as terminals of the tree, while functions are internal nodes. GP iteratively constructs new offspring programs by applying evolutionary operations, crossover and mutation, on the individuals probabilistically chosen based on their fitness; i.e., better individuals are more likely to be chosen to generate more offspring programs. In crossover, one or two programs are created by recombining randomly chosen parts from two selected parent programs. In mutation, a new offspring program is generated by randomly altering a randomly chosen part of a selected program. Some GP systems also support structured solutions and include architecture-altering operations which randomly alter the architecture (e.g., the number of subroutines) of a program to create new offspring programs.

2.2 Differential Evolution

Similar to other evolutionary algorithms, differential evolution is a population-based, derivative-free function optimizer. It usually encodes decision variables as floating-point numbers and manipulates them with simple arithmetic operations such as addition, subtraction, and multiplication. The initial population $\{\mathbf{x}_{i,0} = (x_{1,i,0}, x_{2,i,0}, \ldots, x_{D,i,0}) | i = 1, 2, \ldots, NP\}$ is randomly generated according to a normal or uniform distribution $x_j^{\text{low}} \leq x_{j,i,0} \leq x_j^{\text{up}}$, for $j = 1, 2, \ldots, D$, where NP is the population size, D is the dimension of the problem, and x_j^{low} and x_j^{up} are the upper and lower limits of the j-th component of the vector. After initialization, DE enters a loop of evolutionary operations: mutation, crossover and selection.

Mutation: At each generation g, this operation creates mutation vectors $\mathbf{v}_{i,g}$ based on the current parent population $\{\mathbf{x}_{i,g} = (x_{1,i,g}, x_{2,i,g}, \ldots, x_{D,i,g}) | i = 1, 2, \ldots, NP\}$. The following are different mutation strategies frequently used in the literature,

- 'DE/rand/1'

$$\mathbf{v}_{i,g} = \mathbf{x}_{r0,g} + F_i(\mathbf{x}_{r1,g} - \mathbf{x}_{r2,g}), \tag{2.1}$$

- 'DE/current-to-best/1'

$$\mathbf{v}_{i,g} = \mathbf{x}_{i,g} + F_i(\mathbf{x}_{\text{best},g} - \mathbf{x}_{i,g}) + F_i(\mathbf{x}_{r1,g} - \mathbf{x}_{r2,g}), \tag{2.2}$$

- 'DE/best/1'

$$\mathbf{v}_{i,g} = \mathbf{x}_{\text{best},g} + F_i(\mathbf{x}_{r1,g} - \mathbf{x}_{r2,g}), \tag{2.3}$$

where the indices $r0$, $r1$ and $r2$ are distinct integers uniformly chosen from the set $\{1, 2, \ldots, NP\} \backslash \{i\}$, $\mathbf{x}_{r1,g} - \mathbf{x}_{r2,g}$ is a difference vector to mutate the parent, $\mathbf{x}_{\text{best},g}$ is the best vector at the current generation g, and F_i is the mutation factor which usually ranges on the interval (0, 1+). In classic DE algorithms, $F_i = F$ is a single parameter used for the generation of all mutation vectors, while in many adaptive DE algorithms each individual i is associated with its own mutation factor F_i.

The above mutation strategies can be generalized by implementing multiple difference vectors other than $\mathbf{x}_{r1,g} - \mathbf{x}_{r2,g}$. The resulting strategy is named as 'DE/–/k' depending on the number k of difference vectors adopted.

The mutation operation may generate trial vectors whose components violate the predefined boundary constraints. Possible solutions to tackle this problem include resetting schemes, penalty schemes, etc [2]. A simple method is to set the violating component to be the middle between the violated bound and the corresponding components of the parent individual, i.e.,

$$v_{j,i,g} = (x_j^{\text{low}} + x_{j,i,g})/2, \;\text{if}\; v_{j,i,g} < x_j^{\text{low}}, \tag{2.4}$$

$$v_{j,i,g} = (x_j^{\text{up}} + x_{j,i,g})/2, \;\text{if}\; v_{j,i,g} > x_j^{\text{up}}, \tag{2.5}$$

where $v_{j,i,g}$ and $x_{j,i,g}$ are the j-th components of the mutation vector $\mathbf{v}_{i,g}$ and the parent vector $\mathbf{x}_{i,g}$ at generation g, respectively. This method performs well especially when the optimal solution is located near or on the boundary.

Crossover: After mutation, a 'binomial' crossover operation forms the final trial vector $\mathbf{u}_{i,g} = (u_{1,i,g}, u_{2,i,g}, \ldots, u_{D,i,g})$:

$$u_{j,i,g} = \begin{cases} v_{j,i,g} & \text{if } \text{rand}_j(0,1) \le CR_i \text{ or } j = j_{\text{rand}} \\ x_{j,i,g} & \text{otherwise,} \end{cases} \tag{2.6}$$

where rand$_j$(a,b) is a uniform random number on the interval (a, b] and newly generated for each j, j_{rand} = randint$_i(1, D)$ is an integer randomly chosen from 1 to D and newly generated for each i, and the crossover probability, $CR_i \in [0, 1]$, roughly corresponds to the average fraction of vector components that are inherited from the mutation vector. In classic DE, $CR_i = CR$ is a single parameter that is used to generate all trial vectors, while in many adaptive DE algorithms, each individual i is associated with its own crossover probabilityCR_i.

Selection: The selection operation selects the better one from the parent vector $\mathbf{x}_{i,g}$ and the trial vector $\mathbf{u}_{i,g}$ according to their fitness values $f(\cdot)$. For example, if we have a minimization problem, the selected vector is given by

$$\mathbf{x}_{i,g+1} = \begin{cases} \mathbf{u}_{i,g}, & \text{if } f(\mathbf{u}_{i,g}) < f(\mathbf{x}_{i,g}) \\ \mathbf{x}_{i,g}, & \text{otherwise,} \end{cases} \tag{2.7}$$

and used as a parent vector in the next generation.

The above one-to-one selection procedure is generally kept fixed in different DE algorithms, while the crossover may have variants other than the binomial operation in (2.6). Thus, a DE algorithm is historically named, for example, DE/rand/1/bin connoting its DE/rand/1 mutation strategy and binomial crossover operation.

2.3 Parameter Control

Differential evolution has been shown to be a simple yet powerful evolutionary algorithm for global optimization for many real-world problems [2]. However, like other evolutionary algorithms, DE's performance is quite dependent on the setting of control parameters such as the mutation factor and the crossover probability. Although there are already suggested values of these parameters [1], [9], [10], the interaction between the parameter setting and the convergence performance is still complicated and not completely understood. This is mainly because there is no single parameter value that is suitable for various problems or performs consistently well even at different evolution stages of a single problem.

The trial-and-error method for tuning the control parameters usually requires tedious optimization trials, even for an algorithm (e.g., the classic DE/rand/1/bin) which fixes the parameters throughout the evolutionary search. Motivated by this consideration, different adaptive or self-adaptive mechanisms [16] – [25] have been recently introduced to dynamically update the control parameters without the user's prior knowledge of the problem or interaction during the search process. In addition, parameter adaptation, if well designed, is capable of improving an algorithm's convergence performance.

Adaptation mechanisms proposed in the literature of evolutionary algorithms can be categorized based on how the control parameters are changed. According to the classification scheme introduced by Angeline [47] and Eiben et al. [48], [49], we can interpret three classes of parameter control mechanisms as follows:

1. *Deterministic parameter control*: The control parameter is altered by a deterministic rule without taking into account any feedback from the evolutionary search. One example is the time-dependent change of mutation rates proposed by Holland [5].
2. *Adaptive parameter control*: Feedback from the evolutionary search is used to dynamically change the control parameters. Examples of this class are Rechenberg's "1/5-th rule" [3] and the fuzzy-logic adaptive evolutionary algorithms in [50], [18] and [20]. Several newly proposed DE algorithms, SaDE [21], [81] and jDE [22], [24],[1] and the JADE algorithm described in this monograph belong to this category.
3. *Self-adaptive parameter control*: A method of 'the evolution of evolution' is used to conduct the self-adaptation of control parameters. The parameters are directly associated with individuals and thus themselves undergo mutation and recombination. Since better parameter values tend to generate individuals that are more likely to survive, these values can be propagated to more offspring. Some DE algorithms [16] [17] belong to this category.

Compared to the deterministic method, adaptive or self-adaptive parameter control, if well designed, may enhance the robustness of an algorithm by dynamically adapting the parameters to the characteristic of different fitness landscapes and thus being applicable to various optimization problems without trial and error.

[1] Some of these algorithms are classified as "self-adaptive" in the original papers. However, although each individual in the population is associated with its own control parameters, these parameters themselves do not undergo mutation and crossover in the optimization process. Thus, according to the definition of Eiben et al., these algorithms can be considered as adaptive schemes.

Furthermore, the convergence rate can be improved by adapting the control parameters to be appropriate values at different evolution stages of an optimization problem.

2.4 Multi-objective Optimization

Multi-objective optimization exists commonly in real-world applications such as engineering, financial, and scientific applications [51], because the outcome is directly linked to cost, profit and/or many other criteria that have heavy impacts on performance, safety, environment etc. It is difficult to provide an ultimate comparison among different outcomes in a single dimension, as the involved multiple criteria/objectives are generally competing and non-commensurable. For example, a financial manager needs to take both return and risk into consideration when making an investment decision; an air traffic controller needs to consider both the reduction of system-wide congestion and the satisfaction of each stakeholder's preference.

In mathematical notation, a multi-objective optimization problem can be generally formulated as follows (assume a minimization problem without loss of any generality),

$$\min \mathbf{f}(\mathbf{x}) = (f_1(\mathbf{x}), f_2(\mathbf{x}), \ldots, f_M(\mathbf{x}))^T \tag{2.8}$$

where $\mathbf{x}$ is optimized over a D-dimensional search space and $f_i(\mathbf{x})$ is one of the M objectives to be minimized. There is generally no single solution that simultaneously minimizes all M objectives, which are usually competing and conflicting. As a result, multi-objective optimization problems are usually addressed based on the concepts of Pareto optimum [52], [51]: a Pareto-optimal solution in multi-objective optimization refers to a solution whose corresponding objectives cannot be improved simultaneously; i.e., there is no way to improve any objective without worsening at least one other objective. A formal definition of Pareto optimum is given as follows:

Definition 2.1 (Pareto Optimum). Given the multi-objective optimization defined in (2.8), a solution vector $\mathbf{f} = (f_1, f_2, \ldots, f_M)$ in the objective space is said to dominate another vector $\mathbf{g} = (g_1, g_2, \ldots, g_M)$ if

$$\forall i \in \{1,2,...,M\} : f_i \leq g_i \text{ and } \exists i \in \{1,2,\ldots,M\} : f_i < g_i. \tag{2.9}$$

A solution that is not dominated by any other feasible solutions is said to be Pareto-optimal.

A *Pareto front* is defined as the hypersurface within the objective space that is defined by the set of all Pareto-optimal solutions. That is, the Pareto front is a set of the best "compromise" solutions that cannot be dominated by others (no objective can be improved without making some other objective worse).

The presence of multiple objectives in the optimization of real world applications introduces extra challenges to researchers. More effort is required for these problems beyond traditional techniques like linear and nonlinear programming as well as single-objective evolutionary algorithms. According to the interaction with the decision making process, the multi-objective optimization can be classified as follows [53], [54]:

1. *Prior preference articulation*: The multiple objectives are linearly or nonlinearly aggregated into a scalar function based on a priori preference structure from decision makers. This transforms the problem into a single-objective problem prior to optimization.
2. *Progressive preference articulation*: The optimization is intertwined with the decision-making process when partial preference information is available.
3. *Posterior preference articulation*: A family of tradeoff solutions is found before a decision is made to determine the best solution.

The aggregation method in prior preference articulation is mainly applied in the early development of multi-objective evolutionary algorithms [55]. While the progressive preference articulation is also a current research interest [56], the posterior preference has been gaining significant attention from researchers in proposing multi-objective evolutionary algorithms (MOEAs) to search for tradeoff solutions [35], [57], [58], [59], [60], [61].

2.5 No Free Lunch Theorem and Domain Knowledge Utilization

The No Free Lunch (NFL) theorem for optimization [62] states that the performance of any two search algorithms without incorporating problem specific knowledge is the same if taken over the set of all possible combinatorial optimization problems and that it is important to incorporate problem-specific knowledge into the behavior of the algorithm for performance improvement. Roughly speaking, if an algorithm a_1 performs better than another algorithm a_2, say a random search, for some problems, it will perform worse than random search for other problems to maintain an equally good performance average. It has been further shown [63] that the NFL theorem applies to any subset of functions if and only if that subset is closed under permutation. The NFL theorem therefore implies that we are unable to find a generally best algorithm by benchmarking the entire class of potential problems or a class of problems that compose a set closed under permutation.

The NFL theorem is established based on assumptions of a uniform probability distribution over fitness functions and closure under permutation. It appears that naturally occurring classes of problems are unlikely to be closed under permutation. In [63], it proves that the fraction of non-empty subsets that are closed under permutation rapidly approaches zero as the cardinality of the search space increases, and that constraints on steepness and number of local minima lead to subsets that are not closed under permutation. Thus, it is consistent with intuition that some algorithms generally perform better than others for some naturally occurring classes of structured problems. Nevertheless, it should be born in mind that any potentially optimal algorithm can be considered so only for the limited problems that have been tested. Its performance may be promising for other problems (especially those of similar characteristics) but need to be tested without prior guarantee.

In addition, the NFL theorem implies that an algorithm can outperform another if it is specialized to the structure of a specific problem under consideration [64]. This highlights the need for utilizing domain knowledge in an algorithm (such as EA) to achieve better performance. Indeed, while EAs as well as DE are not problem-specific in nature and have shown great success in many different applications, it has been shown in many practical applications [65] that the performance of EAs can be remarkably improved by incorporating additional problem-specific domain knowledge. According to [65], the approaches of utilizing domain knowledge can be categorized as follows:

1. *Implicit knowledge utilization*: Domain knowledge is utilized for a better representation of the (possibly constrained) search space or a better encoding scheme of the solution (e.g., real-valued or discrete-valued encoding).
2. *Explicit knowledge utilization*: Domain knowledge is utilized to feed high-quality solutions to the initial population or to update control parameters during the optimization process.

In this monograph, we propose a parameter adaptive differential evolution algorithm that is not problem specific in nature. Thus, the utilization of domain knowledge is not considered in the comparison with other state-of-the-art algorithms for a variety of benchmark problems of different characteristics such as multimodality, discontinuity, and high-dimensionality. However, we take into account the domain knowledge in the application of the proposed algorithm for some real-world problems such as the winner determination in combinatorial auction, the transition probability matrix optimization in credit decision-making processes, and the flight planning and optimization in air traffic control systems.

Chapter 3
Theoretical Analysis of Differential Evolution

Differential evolution has proven to be a simple yet efficient optimization approach since its invention by Storn and Price in 1995 [1], [2]. Despite its success in various practical applications, only a few theoretical results [2], [13], [14], [15] have been obtained concerning its stochastic behavior and most of them focus on the mutation and crossover operations while omitting detailed analysis of selection that is immediately related to objective function values and characteristics. The control parameters of DE, both the mutation factor and the crossover probability, are sensitive to the characteristics of different problems and the varied landscapes of a single problem at different evolutionary search stages. Thus, it is necessary to consider the selection operation in investigating the effect of control parameters on the stochastic convergence behavior of differential evolution.

In this chapter, an analytical method is proposed to study the evolutionary stochastic properties of the population of the classic DE for a spherical function model. The effect of crossover probability is relatively clear, i.e., its value mainly depends on the decomposability of the optimization problem [2]. Thus, the classic DE algorithm is analyzed only with its mutation and selection operators. Based on the derived properties of mutation and selection, a Gaussian approximate model of DE is introduced to facilitate mathematical derivations. The evolutionary dynamics and the convergence behavior of DE are investigated based on the analytical formulae derived and their appropriateness is verified by experimental results.

3.1 Introduction

Theoretical analyses have been conducted for evolutionary strategies and genetic algorithms in the last few decades [6], [11], [12]. These results, although generally obtained only for simple fitness models, have proven useful in guiding the development and implementation of existing algorithms and their new variants. For example, Beyer has analyzed several simple ES algorithms [12] using a mathematical model and assuming a spherical fitness landscape (so called sphere model). The analyses are conducted by taking advantage of the fact that the mutation vectors always follow an independent isotropic normal distribution which can be represented

J. Zhang and A.C. Sanderson: Adaptive Differential Evolution, ALO 1, pp. 15–38.
springerlink.com

by a few parameters, i.e., the mean and the variance. The evolution of this algorithm from generation g to $g+1$ is an iterative mapping, and the complete evolutionary dynamics can thus be understood relatively easily. In addition, the small number of representative parameters of the isotropic normal distribution enables a tractable mathematical derivation. As a result, theoretical results are obtained in closed form or integral form with the standard normal cumulative distribution function (cdf) $\Phi_0(\cdot)$ appearing in the integrand. It is worth noting that the same approach has been used in [66], [67] to study the performance of the (μ,λ)-ES on the noisy sphere model.

Analyses of other elegant evolutionary algorithms are more complicated. However, if the parent and/or mutation population always follows the same family of distributions which has a simple analytical form, mathematical analyses may still be manageable for simple fitness models (e.g., hyper-plane, sphere, ridge models [68], [69]). Nevertheless, the problem is intractable if the population distribution is difficult to formulate in a simple concise form. One example is the family of $(\mu+\lambda)$ ES [11] for which the distribution of the union of parent and mutation populations is necessary for analytical derivations. The distribution is difficult to represent because of the correlation between the two populations. Thus, these strategies, except the simpler form of $(1+\lambda)$ [70] , have not been theoretically analyzed in the literature so far. Another example is the ESs (e.g., CORR-ES and CMA-ES [71]) which adapt the full $D\times D$ covariance matrix of the population, where D is the dimension of each individual vector. The behavior of these strategies is dependent on up to $D(D+1)/2$ parameters even if a normal distribution is assumed. Thus, theoretical analyses are again difficult.

The properties of the population distribution also do not hold for the differential evolution. In contrast to ESs and GAs, the parent and mutation populations of DE are not generated according to a predefined family of distributions. The random process induced by DE is too complex to be rigorously tackled by a mathematical analysis: the distribution varies from generation to generation as a result of the problem-dependent selection operation. It is difficult to obtain an analytical expression for such a distribution, even if the initial population is generated according to an isotropic normal distribution. Indeed, to the author's knowledge, the DE performance on the sphere model has been analyzed only by empirical studies such as [72]. It motivates us to make an appropriate approximation on the population distribution to facilitate a tractable mathematical analysis.

In this chapter, we consider an *a*pproximate model of DE (ADE model) which differs from the *s*tandard DE/rand/*k*/bin algorithm without crossover (SDE model) only in its assumption of a normal distribution of the parent population. To be specific, the first- and second-order statistics of the practical parent population are used as the mean and variance of the population's normal distribution in the approximate model. All other components of ADE and SDE are identical, including the differential mutation and one-to-one selection. In this manner, an analytical derivation becomes possible for differential evolution. Similar to the analysis of ES, the final results are obtained for DE in closed form or integral form with $\Phi_0(\cdot)$ appearing in the integrand.

3.2 Properties of Differential Evolution

As there are many variants of differential evolution as described in Chapter 2.2, we focus on the classic DE/rand/*k*/bin and ignore the crossover operation for the simplicity of derivation. We develop stochastic properties for the mutation and selection operations and then introduce an approximate model of DE based on these properties.

For the sake of notational convenience, the notations used in this chapter are slightly different from those in Chapter 2.2 and other chapters: the mutation vectors are denoted as $\mathbf{y}$ (instead of $\mathbf{u}$) and the new parent vectors as $\mathbf{z}$ (instead of $\mathbf{x}_{g+1}$).

3.2.1 DE/rand/k/bin without Crossover

In DE/rand/*k*/bin, the initial population $\{\mathbf{x}_{i,0} = (x_{1,i,0}, x_{2,i,0}, \ldots, x_{D,i,0}) | i = 1, 2, \ldots, NP\}$ is created by random sampling from a certain distribution, e.g., an isotropic normal distribution. Then the algorithm enters a loop of mutation and selection. The operation at generation g is formulated as follows.

Mutation: Each mutation vector $\mathbf{y}_{i,g}$ is generated by mutating a parent vector $\mathbf{x}_{r_0,g}$ with the scaled sum of k parent-vector differences $\mathbf{x}_{r_1^l,g} - \mathbf{x}_{r_2^l,g}$, $l = 1, 2, \ldots, k$. That is,

$$\mathbf{y}_{i,g} = \mathbf{x}_{r0,g} + F \sum_{l=1}^{k} (\mathbf{x}_{r_1^l,g} - \mathbf{x}_{r_2^l,g}), i = 1, 2, \ldots, NP, \tag{3.1}$$

where F is the mutation factor fixed throughout the optimization. The indices $r0$, r_1^l and r_2^l are distinct integers uniformly chosen from the set $\{1, 2, \ldots, NP\} \backslash \{i\}$.

Selection: Select the better one, denoted as $\mathbf{z}_{i,g}$, from the parent vector $\mathbf{x}_{i,g}$ and the mutation vector $\mathbf{y}_{i,g}$ according to their fitness values. That is,

$$\mathbf{z}_{i,g} = \begin{cases} \mathbf{y}_{i,g}, \text{ if} f(\mathbf{y}_{i,g}) < f(\mathbf{x}_{i,g}) \\ \mathbf{x}_{i,g}, \text{ otherwise}, \end{cases} \tag{3.2}$$

where $f(\cdot)$ is the fitness function to be minimized. This vector is used as a parent in the next generation, i.e., $\mathbf{x}_{i,g+1} = \mathbf{z}_{i,g}$.

Differential evolution (without crossover) described above satisfies the following property [2]:

Property 3.1. The differential evolution algorithm defined by (3.1) and (3.2) is rotationally invariant; i.e., the mutation and selection operations are independent of the orientation of the coordinate system in which the fitness function is evaluated.

For notational simplicity, we omit the subscript g in the rest of the chapter if no confusion is possible.

3.2.2 Properties of Mutation

Suppose all parent vectors $\mathbf{x}_i$'s are independently generated according to an identical probability density distribution (pdf) $p_{\mathbf{x}}(\mathbf{x})$. Then, it is easy to prove that $\mathbf{x}_i$, $\mathbf{x}_{r0}$,

$\mathbf{x}_{r_1^l}$ and $\mathbf{x}_{r_2^l}$ are i.i.d. random vectors given that i, r_0, r_1^l and r_2^l are distinct indices. Thus, if $p_\mathbf{x}(\mathbf{x})$ has mean $\mu_\mathbf{x}$ and covariance $\mathbf{C}_x$, then $\mathbf{y}_i$, as a linear combination of i.i.d. variables with pdf $p_\mathbf{x}(\mathbf{x})$, follows a distribution $p_\mathbf{y}(\mathbf{y})$ of mean $\mu_\mathbf{y} = \mu_\mathbf{x}$ and covariance $\mathrm{C}_\mathbf{y} = (1+2kF^2)\mathrm{C}_\mathbf{x}$. This property is summarized as follows:

Property 3.2. If the parent vectors $\mathbf{x}_i$'s are i.i.d. with mean $\mu_\mathbf{x}$ and covariance $\mathrm{C}_\mathbf{x}$, the mutation vector $\mathbf{y}_i$ generated according to (3.1) is independent of $\mathbf{x}_i$ and its mean and covariance are given by

$$\mu_\mathbf{y} = \mu_\mathbf{x} \text{ and } \mathrm{C}_\mathbf{y} = (1+2kF^2)\mathrm{C}_\mathbf{x}, \tag{3.3}$$

respectively. In addition, if $\mathbf{x}_i$ is normally distributed, $p_\mathbf{x} = \mathcal{N}(\mu_\mathbf{x}, \mathrm{C}_\mathbf{x})$, $\mathbf{y}_i$ is normally distributed as well and $p_\mathbf{y}(\mathbf{y}) = \mathcal{N}(\mu_\mathbf{y}, \mathrm{C}_\mathbf{y})$.

It is worth noting that the distributions concerned in this chapter are ensemble distributions , i.e., the distributions that are followed by the vectors $\mathbf{x}_i$, $\mathbf{y}_i$ or $\mathbf{z}_i$ when considering the ensemble of random realizations of the initialization, mutations, and selections. This should be distinguished from the sample distribution after a certain realization of initialization, mutations and selections. Note that while $p_\mathbf{x}(\mathbf{x})$, $p_\mathbf{y}(\mathbf{y})$ and $p_\mathbf{z}(\mathbf{z})$ are continuous distribution functions, the sample distribution function is merely the sum of *NP* delta functions, each centered at a sample vector after the initialization, mutation and selection.

As it becomes clear below, the ensemble distribution has desirable properties (which a sample distribution usually lacks) to facilitate our mathematical derivation. However, it is worth noting that the result derived based on these ensemble properties can well predict the behavior of a single DE experiment only when the population size is sufficiently large (usually larger than the common values $NP = 3D \sim 5D$).

For simplicity, the distribution discussed in the rest of this chapter implicitly refers to the ensemble distribution, if not otherwise explained.

3.2.3 *Properties of Selection*

In contrast to the mutation, the selection of DE is dependent on the landscape of the fitness function. Consider the (hyper-)sphere model [12] in which the fitness function $f(\mathbf{x})$ is symmetric around its center; i.e., the isometric surface "$f(\mathbf{x})$ = const" draws concentric (hyper-)sphere shells. For notational convenience, we use the origin O as the center of symmetry and denote r_x as the distance of a vector $\mathbf{x}$ to the origin (i.e., $r_x = \| \mathbf{x} \|$, the length of $\mathbf{x}$). The fitness function $f(\mathbf{x})$ is a function of type $Q(r_x)$ which only depends on r_x. It is assumed that $Q(r_x)$ is a strongly monotonic function having its optimum at $r_x = 0$. In this case, the selection procedure in (3.2) can be rewritten as

$$\mathbf{z}_i = \begin{cases} \mathbf{x}_i, & r_x^2 \le r_y^2 \\ \mathbf{y}_i, & r_x^2 > r_y^2. \end{cases} \tag{3.4}$$

Assume the initial population is generated according to an isotropic normal distribution. This distribution satisfies the *property of rotational symmetry*: a distribution is said to be rotationally symmetric if it is isotropic in all directions orthogonal to a given axis across the origin. According to the geometric symmetry of the sphere model around the origin (the optimum), we can assume without loss of generality that the mean of the initial distribution is located on the first coordinate axis. We can draw the following conclusion by resorting to the geometric symmetry of the sphere model and the rotational invariance in Property 3.1 (The proof is lengthy but standard. See the appendix in Section 3.7.1 for details):

Property 3.3. If the distribution of the initial parent population $\mathbf{x}_{i,0}$ is rotationally symmetric around the first coordinate axis, so are the distributions of sequential populations $\mathbf{y}_{i,g}$, $\mathbf{z}_{i,g}$ and $\mathbf{x}_{i,g+1}$ and their respective sample mean $\bar{\mathbf{y}}_g$, $\bar{\mathbf{z}}_g$ and $\bar{\mathbf{x}}_{g+1}$, $g = 0, 1, 2, \ldots$

We next restrict attention to the population $\{\mathbf{z}_i\}$. Denote $z_{j,i}$ as the j-th component of each individual $\mathbf{z}_i$ and $\{z_{j,i}\}$ as the set of the j-th components of all individuals in $\{\mathbf{z}_i\}$. The (marginal) distribution of $z_{j,i}$ can be obtained by integrating the pdf $p^g_{\mathbf{z}}(\mathbf{z})$ of $\mathbf{z}_i$ with respect to all other components. According to Property 3.3, $p^g_{\mathbf{z}}(\mathbf{z})$ is rotationally symmetric around the first coordinate axis, i.e., it is isotropic with respect to its second to last components. Thus, it is easy to prove the following property for $z_{j,i}$ (Similar to Property 3.3, the proof is very standard and omitted here):

Property 3.4. If the assumption in Property 3.3 holds, the component $z_{j,i}$ ($j \neq 1$) of $\mathbf{z}_i$ follows an identical distribution which is symmetric around zero, and so are the sample mean of $\{z_{j,i}\}$, $j \neq 1$. In addition, the sample variance of $z_{j,i}$, $j \neq 1$, follows an identical distribution.

According to Property 3.4, the average sample mean $E(\bar{z}_j)$ of $\{z_{j,i}\}$ is equal to zero if $j \neq 1$. Thus, the average sample mean of $\{\mathbf{z}_i\}$ is of the form

$$E(\bar{\mathbf{z}}) = (E(\bar{z}_1), E(\bar{z}_2), \cdots, E(\bar{z}_D)) = (E(\bar{z}_1), 0, \cdots, 0). \tag{3.5}$$

In addition, the sample variance of $z_{j,i}$ are identical for $j \neq 1$. Denote the average sample variance of $\{z_{j,i}\}$ as $E(s^2_{z1})$ for $j = 1$ and $E(s^2_{z2})$ for $j \neq 1$, respectively.

3.3 An Approximate Model of DE

In Sections 3.2.1 – 3.2.3, we have derived some properties for the population distribution of DE for the sphere model. However, the exact formula of the distribution is still too complex to be rigorously tackled by a mathematical analysis, even though the initial population is created from an isotropic normal distribution. Thus, for the sake of tractable mathematical derivations, it is necessary to consider an approximate model of DE (ADE model) at the cost of moderate accuracy. The ADE model differs from the standard DE (SDE model) only in that it assumes the distribution $p^g_{\mathbf{x}}(\mathbf{x})$ at each generation belongs to a distribution family, which has a concise analytical form. The family of normal distributions is considered because of its simple expression and its approximation to the situation in various experiments of DE on

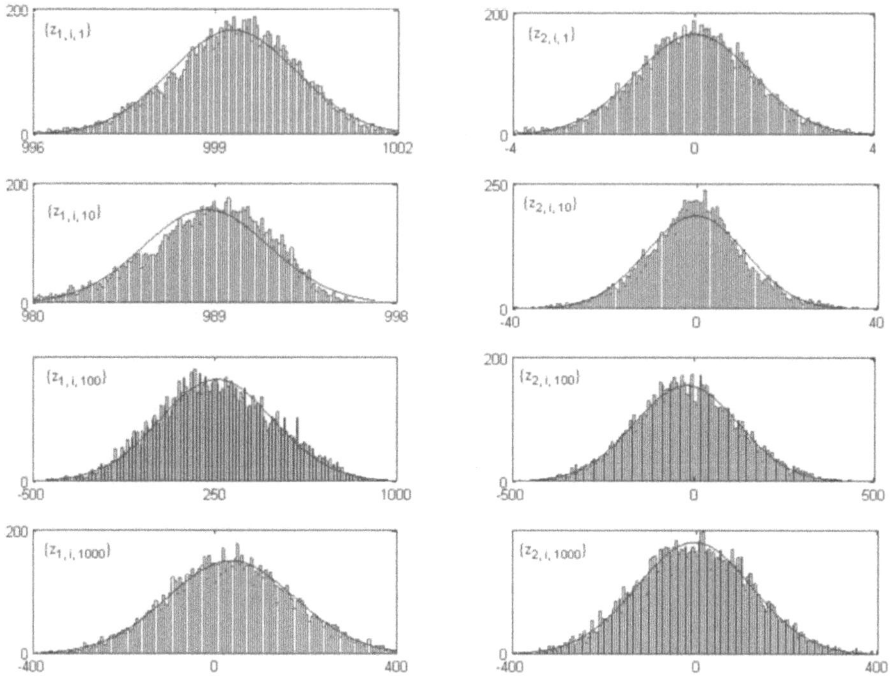

Fig. 3.1 Histogram of $\{z_{j,i,g}\}$ in a DE experiment with $R_0 = 10^3$ and $\sigma_{x1,0} = \sigma_{x2,0} = 1$ ($k = 1$, $F = 0.8$, $D = 30$, $NP = 10000$). The smooth curve shows the scaled pdf of a normal distribution whose mean and variance are identical to the sample mean and sample variance of $\{z_{j,i,g}\}$, respectively

the sphere model (See Figure 3.1 for an example). All other components of ADE, including the differential mutation and the one-to-one selection in (3.1) and (3.2), are assumed to be identical to those of SDE.

To fulfill the approximation of the normal distribution, we assume that, instead of simply setting $\mathbf{x}_{i,g} = \mathbf{z}_{i,g}$ in SDE, the parent vectors $\mathbf{x}_{i,g+1}$ are generated in ADE according to the normal distribution whose parameters conform to the first- and second-order statistics of $\{\mathbf{z}_{i,g}\}$. For the convenience of mathematical analysis, we consider a normal distribution

$$\mathcal{N}(R, \sigma_1^2, \sigma_2^2) := \mathcal{N}(\mu_{\mathbf{x}}, \mathrm{C}_{\mathbf{x}}), \tag{3.6}$$

with mean $\mu_{\mathbf{x}} = (R, 0, 0, \ldots, 0)$, variances $\mathrm{C}_{1,1} = \sigma_1^2$, $\mathrm{C}_{i,i} = \sigma_2^2$ for $i \neq 1$, and covariance $\mathrm{C}_{i,j} = 0$ for $i \neq j$. Note that this distribution satisfies the stochastic properties developed above, including the property of rotational symmetry. Denote the distribution of $\mathbf{x}_{i,g+1}$ as $p_{\mathbf{x}}^{g+1}(x) = \mathcal{N}(R_g, \sigma_{x1,g+1}^2, \sigma_{x2,g+1}^2)$. The two variance parameters $\sigma_{x1,g+1}^2$ and $\sigma_{x2,g+1}^2$ are set to be the corresponding average sample variances $E(s_{z1,g}^2)$ and $E(s_{z2,g}^2)$ of $\{\mathbf{z}_{i,g}\}$. The parameter R_g represents the distance of the mean

$(R_g, 0, \ldots, 0)$ to the optimum (the origin). It is set to be the average distance $E(r_{\bar{z}})$ of the sample mean $\bar{\mathbf{z}}_g$ of $\{\mathbf{z}_{i,g}\}$ to the optimum. That is, we have

$$\begin{cases} R_{g+1} = E(r_{\bar{z}}) \\ \sigma^2_{x1,g+1} = E(s^2_{z1,g}) \\ \sigma^2_{x2,g+1} = E(s^2_{z2,g}) \end{cases} \tag{3.7}$$

The equations in (3.7) represent an iterative mapping based on which the analysis of the stochastic behavior from generation g to $g+1$ can be used to understand the complete evolutionary dynamics of the ADE model and thus provides valuable information on the SDE model as well.

3.4 Analyses of the Evolution Process of DE

In this section, we make use of the approximate model of DE to investigate its evolution behavior from one generation to the next. To be specific, we calculate the right-hand sides of the iterative mapping (3.7) based on the stochastic characteristics of the parent population at generation g. These are set to be the distribution parameters at the next generation according to (3.7).

3.4.1 Mathematical Formulation of DE

The selection procedure of DE is illustrated in Figure 3.2 in the case of the sphere model. Each pair of parent vector $\mathbf{x} = (x_1, x_2, \ldots, x_D)$ and mutation vector $\mathbf{y} = (y_1, y_2, \ldots, y_D)$ is compared in terms of their fitness, i.e., their (squared) distances to the origin. The better one is selected and denoted as $\mathbf{z} = (z_1, z_2, \ldots, z_D)$. The vector $\mathbf{x}$ is assumed to be generated independently according to a normal pdf $p_{\mathbf{x}}(\mathbf{x}) = \mathcal{N}(R, \sigma^2_{x1}, \sigma^2_{x2})$. According to Property 2, the mutation vector $\mathbf{y}$ is independent of $\mathbf{x}$ and is normally distributed with pdf $p_{\mathbf{y}}(\mathbf{y}) = \mathcal{N}(R, \sigma^2_{y1}, \sigma^2_{y2})$, where $\sigma^2_{yi} = (1 + 2kF^2)\sigma^2_{xi}$, $i = 1, 2$.

Denote the first coordinate of $\mathbf{x}$ as $x_1 = R - x$ and the summed square of other coordinates as $h^2_x = x^2_2 + x^2_3 + \cdots + x^2_D$. The squared distance of $\mathbf{x}$ to the origin is given by $r^2_x = h^2_x + (R - x)^2$. Similar notations are introduced for $\mathbf{y}$ and $\mathbf{z}$, including h^2_y, r^2_y, h^2_z, r^2_z and $y_1 = R - y$, $z_1 = R - z$. For $\mathbf{x}$, we have $x \sim \mathcal{N}(0, \sigma^2_{x1})$ and $x_i \sim \mathcal{N}(0, \sigma^2_{x2})$ for $i \neq 1$. Thus, the summed square h^2_x follows a chi-squared distribution of degree $D - 1$. If $D >> 1$, say $D \geq 30$, it can be well approximated as a normal random variable with mean $\tilde{D}\sigma^2_{x2}$ and variance $2\tilde{D}\sigma^4_{x2}$, where $\tilde{D} = D - 1$. That is, $h^2_x \sim \mathcal{N}(\tilde{D}\sigma^2_{x2}, 2\tilde{D}\sigma^4_{x2})$. Similarly, we have $y \sim \mathcal{N}(0, \sigma^2_{y1})$ and $h^2_y \sim \mathcal{N}(\tilde{D}\sigma^2_{y2}, 2\tilde{D}\sigma^4_{y2})$.

The distribution of $\mathbf{z}$ is difficult to obtain analytically. However, according to Property 3.4, we know that z_i $(i \neq 1)$ follows an identical distribution which is symmetric around zero (i.e., $E(z_i) = 0$, $i \neq 1$). Denote the variance of z_i as σ^2_{z1} and σ^2_{z2} for $i = 1$ and $i \neq 1$, respectively. The variance of z, $z = R - z_1$, is equal to σ^2_{z1} as well. In addition, $E(z_1) = R - E(z)$.

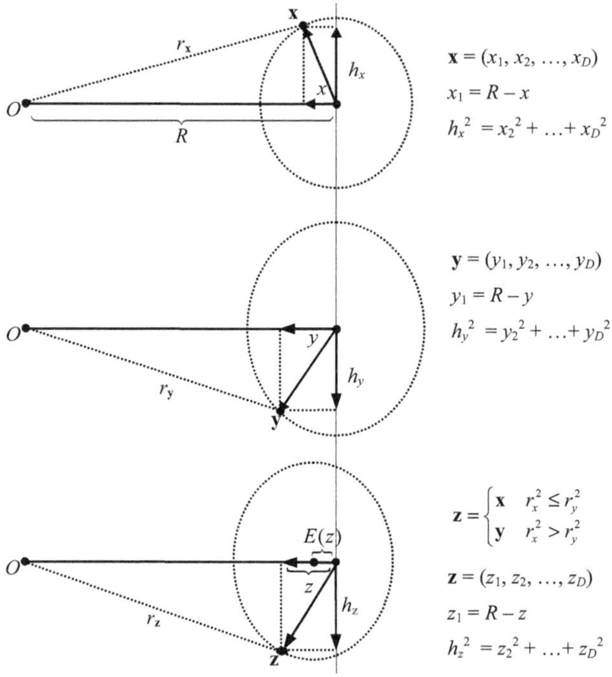

Fig. 3.2 An illustration of the selection in SDE and ADE. The point at the center of the ellipses is the population mean

In the following subsections, we derive analytical expressions for $E(z)$, σ_{z1}^2 and σ_{z2}^2. The average sample mean and sample variance of $\{\mathbf{z}\}$ are then calculated to obtain the right hand sides of the iterative mapping (3.7). Before proceeding with detailed derivations, it is convenient to express the selection procedure (3.4) in an equivalent form:

$$z = \begin{cases} x, \ h_x^2 + (R-x)^2 \le h_y^2 + (R-y)^2 \\ y, \ h_x^2 + (R-x)^2 > h_y^2 + (R-y)^2, \end{cases} \tag{3.8}$$

and

$$h_z^2 = \begin{cases} h_x^2, \ h_x^2 + (R-x)^2 \le h_y^2 + (R-y)^2 \\ h_y^2, \ h_x^2 + (R-x)^2 > h_y^2 + (R-y)^2, \end{cases} \tag{3.9}$$

where

$$\begin{gathered} x \sim \mathscr{N}(0, \sigma_{x1}^2), y \sim \mathscr{N}(0, \sigma_{y1}^2), \\ h_x^2 \sim \mathscr{N}(\tilde{D}\sigma_{x2}^2, 2\tilde{D}\sigma_{x2}^4), h_y^2 \sim \mathscr{N}(\tilde{D}\sigma_{y2}^2, 2\tilde{D}\sigma_{y2}^4). \end{gathered} \tag{3.10}$$

3.4.2 *Calculation of E(z)*

Both r_x^2 and r_y^2 are linear combinations of central and non-central chi-squared random variables. There is no analytical expression for the pdf of this type of random variables, although efficient numerical methods are available in the literature [73], [74]. In view of this, we consider an approximation of r_x^2,

$$r_x^2 = h_x^2 + (R-x)^2 \approx h_x^2 + R^2 - 2Rx, \tag{3.11}$$

by neglecting the term x^2 when either $x << R$ or $x^2 << h_x^2$. This is the method generally used in the analyses of ES [12] and it works well in those situations. Its validity still needs further investigation in our case. The advantage of (3.11) is that every term in the expansion of r_x^2 is either constant or normally distributed, making it possible to derive a closed-form expression for $E(z)$.

By applying a similar approximation to r_y^2, we can rewrite (3.8) as follows:

$$z = \begin{cases} x, \; h_x^2 - h_y^2 + 2Ry \le 2Rx \\ y, \; h_y^2 - h_x^2 + 2Rx < 2Ry. \end{cases} \tag{3.12}$$

Denote $p_x(x)$ and $p_y(y)$ as the pdf of x and y, respectively. The pdf of z is given by

$$p_z(z) = p_x(z)\Pr\{h_x^2 - h_y^2 + 2Ry \le 2Rz\} + p_y(z)\Pr\{h_y^2 - h_x^2 + 2Rx < 2Rz\}. \tag{3.13}$$

Both $h_x^2 - h_y^2 + 2Ry$ and $h_y^2 - h_x^2 + 2Rx$ are normal random variables whose mean and variance can be easily obtained according to (3.10). Thus, we can write (3.13) as follows:

$$\begin{aligned} p_z(z) = & \frac{1}{\sigma_{x1}}\phi_0\left(\frac{z}{\sigma_{x1}}\right)\Phi_0\left(\frac{2Rz-\tilde{D}(\sigma_{x2}^2-\sigma_{y2}^2)}{\sqrt{4R^2\sigma_{y1}^2+2\tilde{D}(\sigma_{x2}^4+\sigma_{y2}^4)}}\right) \\ & +\frac{1}{\sigma_{y1}}\phi_0\left(\frac{z}{\sigma_{y1}}\right)\Phi_0\left(\frac{2Rz-\tilde{D}(\sigma_{y2}^2-\sigma_{x2}^2)}{\sqrt{4R^2\sigma_{x1}^2+2\tilde{D}(\sigma_{x2}^4+\sigma_{y2}^4)}}\right), \end{aligned} \tag{3.14}$$

where $\phi_0(\cdot)$ and $\Phi_0(\cdot)$ denotes the pdf and cdf of a standard normal distribution, respectively. Thus, the expectation of z is given by

$$\begin{aligned} E(z) = & \int_{-\infty}^{\infty}\frac{z}{\sigma_{x1}}\phi_0\left(\frac{z}{\sigma_{x1}}\right)\Phi_0\left(\frac{z-\tilde{D}(\sigma_{x2}^2-\sigma_{y2}^2)/2R}{\sqrt{\sigma_{y1}^2+\tilde{D}(\sigma_{x2}^4+\sigma_{y2}^4)/2R^2}}\right)dz \\ & +\int_{-\infty}^{\infty}\frac{z}{\sigma_{y1}}\phi_0\left(\frac{z}{\sigma_{y1}}\right)\Phi_0\left(\frac{z-\tilde{D}(\sigma_{y2}^2-\sigma_{x2}^2)/2R}{\sqrt{\sigma_{x1}^2+\tilde{D}(\sigma_{x2}^4+\sigma_{y2}^4)/2R^2}}\right)dz \end{aligned} \tag{3.15}$$

This expression can be simplified by using the equation (See the appendix in Section 3.7.2 for proof)

$$\int_{-\infty}^{\infty}\frac{x}{\sigma_x}\phi_0\left(\frac{x}{\sigma_x}\right)\Phi_0\left(\frac{x-\mu_y}{\sigma_y}\right)dx = \frac{\sigma_x^2}{\sqrt{2\pi}\sqrt{\sigma_x^2+\sigma_y^2}}\exp\left(-\frac{\mu_y^2}{2(\sigma_x^2+\sigma_y^2)}\right), \quad \sigma_y > 0. \tag{3.16}$$

By simple manipulation, we have

$$E(z) = \frac{2R(\sigma_{x1}^2+\sigma_{y1}^2)}{\sqrt{2\pi}\sqrt{4R^2(\sigma_{x1}^2+\sigma_{y1}^2)+2\tilde{D}(\sigma_{x2}^4+\sigma_{y2}^4)}}\exp\left(-\frac{\tilde{D}^2(\sigma_{x2}^2-\sigma_{y2}^2)^2}{8R^2(\sigma_{x1}^2+\sigma_{y1}^2)+4\tilde{D}(\sigma_{x2}^4+\sigma_{y2}^4)}\right). \tag{3.17}$$

As a special case, if the hyper-plane model (i.e., sphere model with $R/\sigma_{x1}, R/\sigma_{x2} \to \infty$) is considered, we have

$$E(z) = \frac{1}{\sqrt{2\pi}}\sqrt{\sigma_{x1}^2 + \sigma_{y1}^2}. \tag{3.18}$$

Now, it is necessary to verify the effectiveness of (3.17) by comparing it with the result obtained without the approximation of (3.11). As expected, the more accurate result has a rather complex expression.

Let us consider an intermediate variable $z - y$. According to (3.8), we have

$$z - y = \begin{cases} x - y, & h_x^2 - h_y^2 + (x-y)(x+y-2R) \le 0 \\ 0, & h_x^2 - h_y^2 + (x-y)(x+y-2R) > 0. \end{cases} \tag{3.19}$$

The pdf of $z - y$ is given by

$$\begin{aligned} p_{z-y}(w) &= \delta(w)\Pr\{h_x^2 - h_y^2 + (x-y)(x+y-2R) > 0\} \\ &\quad + p_{x-y}(w)\Pr\{h_x^2 - h_y^2 + w(x+y-2R) \le 0 | x-y = w\}, \end{aligned} \tag{3.20}$$

where $\delta(\cdot)$ is Dirac's delta function.

The first term of (3.20) is not equal to zero only when $w = 0$; thus it can be ignored as far as the mean of $z - y$ is concerned. The second term is complicated because it involves a conditional probability. This probability can be calculated using the following transformation. Given $x - y = w$, we have

$$x + y = \frac{2(\sigma_{y1}^2 x + \sigma_{x1}^2 y)}{\sigma_{x1}^2 + \sigma_{y1}^2} - \frac{w(\sigma_{y1}^2 - \sigma_{x1}^2)}{\sigma_{x1}^2 + \sigma_{y1}^2}, \tag{3.21}$$

by simple manipulation.

The second term of (3.21) is a constant while the first one, the scaled $\sigma_{y1}^2 x + \sigma_{x1}^2 y$, is independent of $x - y$, as can be easily verified [75, 5.1.2]. Thus, by the substitution of (3.21), the conditional probability in (3.20) can be written as:

$$\begin{aligned} &\Pr\{h_x^2 - h_y^2 + w(x+y-2R) \le 0 | x - y = w\} \\ &= \Pr\{\underbrace{h_x^2 - h_y^2 + \frac{2w(\sigma_{y1}^2 x + \sigma_{x1}^2 y)}{\sigma_{x1}^2 + \sigma_{y1}^2}}_{v} \le \frac{w^2(\sigma_{y1}^2 - \sigma_{x1}^2)}{\sigma_{x1}^2 + \sigma_{y1}^2} + 2wR\}, \end{aligned} \tag{3.22}$$

where the variable v is independent of $x - y$ and follows a normal distribution. With the last equation and (3.20), we can obtain the mean $E(z) = E(z - y)$ as follows:

$$E(z) = \int_{-\infty}^{\infty} \frac{w}{\sqrt{\sigma_{x1}^2 + \sigma_{y1}^2}} \phi_0\left(\frac{w}{\sqrt{\sigma_{x1}^2 + \sigma_{y1}^2}}\right) \Phi_0\left(\frac{2wR + \frac{w^2(\sigma_{y1}^2 - \sigma_{x1}^2)}{\sigma_{x1}^2 + \sigma_{y1}^2} - \tilde{D}(\sigma_{x2}^2 - \sigma_{y2}^2)}{\sqrt{2\tilde{D}(\sigma_{y2}^4 + \sigma_{x2}^4) + \frac{4w^2\sigma_{x1}^2\sigma_{y1}^2}{\sigma_{x1}^2 + \sigma_{y1}^2}}}\right) dw, \tag{3.23}$$

where the means and variances of $x-y$ and v, calculated according to (3.10), are used in the arguments of $\phi_0(\cdot)$ and $\Phi_0(\cdot)$.

A substitution of $t = w/\sqrt{\sigma_{x1}^2+\sigma_{y1}^2}$ into (3.23) yields:

$$E(z) = \frac{\sqrt{\sigma_{x1}^2+\sigma_{y1}^2}}{\sqrt{2\pi}} \int_{-\infty}^{\infty} t e^{-t^2/2} \Phi_0 \left(\frac{(\sigma_{y1}^2-\sigma_{x1}^2)t^2+2\sqrt{\sigma_{x1}^2+\sigma_{y1}^2}Rt-\tilde{D}(\sigma_{x2}^2-\sigma_{y2}^2)}{\sqrt{4\sigma_{x1}^2\sigma_{y1}^2t^2+2\tilde{D}(\sigma_{y2}^4+\sigma_{x2}^4)}} \right) dt. \quad (3.24)$$

This expression is similar to the result obtained for ES analyses [12] in the sense that a standard normal cdf $\Phi_0(\cdot)$ appears in the integrand. Further simplification is hard to do because $\sigma_{x1} \neq \sigma_{y1}$ and $\sigma_{x2} \neq \sigma_{y2}$ in DE. However, the integral of (3.24) can be truncated to an interval $[-5,5]$ in practice without much loss of accuracy by noting that the integrand approaches zero super-exponentially fast as the magnitude of t increases.

A simple comparison of (3.17) and (3.24) is illustrated in Figure 3.3. It is clear that Eq. (3.24) has a good agreement with the simulation result. The slight discrepancy is due to the approximation of h_x^2 and h_y^2 as normal variables in (3.10). On the contrary, the accuracy of Eq. (3.17) is acceptable only when σ_{x1} is much less than R (i.e., $x << R$) or σ_{x1} is less than σ_{x2} (i.e., $x^2 << h_x^2$). Thus, benefiting from its

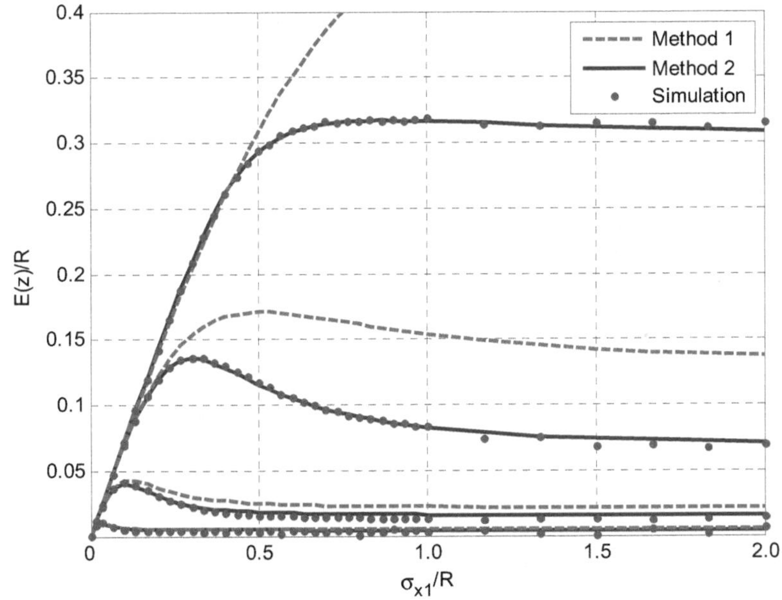

Fig. 3.3 A comparison of two analytical methods to calculate $E(z)$ ($k=1, F=0.8, D=30, NP=600$). From top to bottom, the curves shows the values for σ_{x1}/σ_{x2} = 4, 2, 1 and 1/2. Method 1: Eq. (3.17); Method 2: Eq. (3.24). All the data points are obtained according to one-generation experiments (500 runs)

closed-form expression, Eq. (3.17) is suitable for asymptotic analysis for $R \to \infty$ (i.e., Eq.(3.18) is valid) or $D \to \infty$ (i.e., $x^2 << h_x^2$).

3.4.3 Calculation of σ_{z1}

The standard deviation σ_{z1} of z can be obtained based on the equation $\sigma_{z1}^2 = E(z^2) - E(z)^2$. In what follows, we focus on the calculation of $E(z^2)$, since $E(z)$ has already been obtained.

A straightforward method makes use of the pdf derived in (3.14). Similar to (3.17), however, this leads to inaccurate results when σ_{x1} is close to or greater than R. An alternative approximate method is needed. Instead of (3.12), let us consider the original definition of z in (3.8) and formulate its pdf as follows:

$$\begin{aligned} p_z(z) = \; & p_x(z)\Pr\{h_x^2 - h_y^2 - (R-y)^2 \le -(R-z)^2\} \\ & + p_y(z)\Pr\{h_y^2 - h_x^2 - (R-x)^2 < -(R-z)^2\}. \end{aligned} \tag{3.25}$$

Both $w_1 := h_x^2 - h_y^2 - (R-y)^2$ and $w_2 := h_y^2 - h_x^2 - (R-x)^2$ are linear combinations of central and non-central chi-squared random variables. There is no closed-form analytical expression for the pdf of w_1 or w_2. However, an approximate method can be considered which is built on the expansion of a normal distribution in terms of orthogonal Hermite polynomials. The pdf of a random variable w can be approximated as [12], [76]:

$$p_w(w) = \frac{1}{\sqrt{2\pi}\sigma}\exp\left(-\frac{(w-\mu)^2}{2\sigma^2}\right)\left[1 + \frac{\kappa_3}{3!\sigma^3}H_3\left(\frac{w-\mu}{\sigma}\right) + \frac{\kappa_4}{4!\sigma^4}H_4\left(\frac{w-\mu}{\sigma}\right) + \cdots\right], \tag{3.26}$$

where μ, σ^2, κ_3 and κ_4 are the mean, variance and the third and fourth cumulants of w, respectively, and the two Hermite polynomials are $H_3(x) = x^3 - 3x$ and $H_4(x) = x^4 - 6x^2 + 3$. The bracketed terms in (3.26) are corrections to the normal distribution, of which the first two are generally used in practice and thus expressed in (3.26). If the variable is nearly normally distributed, a few correction terms are generally sufficient to provide a good approximation. Indeed, in the case of w_1 and w_2, experimental results show that the benefit of correction terms is trivial and sufficient accuracy can be achieved by simply approximating them as normally distributed (i.e., with no correction term):

$$p_w(w) \approx \frac{1}{\sqrt{2\pi}\sigma}\exp\left(-\frac{(w-\mu)^2}{2\sigma^2}\right). \tag{3.27}$$

Note that this is similar to the normal approximation of h_x^2 and h_y^2 which is based on the central limit theorem.

Before proceeding with the derivation of $E(z^2)$, it is worth some explanations of the normal approximation of w_1 and w_2. Take w_2 as an example. If $\sigma_{x1} << R$ and thus $x << R$, we have $w_2 \approx h_y^2 - h_x^2 + 2Rx - R^2$ which is nearly normally distributed. This is actually the case of (3.11). Then, let us consider the case when $\sigma_{x1} \approx R$ or $\sigma_{x1} >> R$. In this aspect, we shall note that DE generally operates with $\sigma_{x2} \ge \sigma_{x1}$

or $\sigma_{x2} \approx \sigma_{x1}$ on the sphere model (Intuitively, this is because the selection imposes more pressure on the direction towards the optimum than on any other orthogonal direction). Thus, $\sigma_{x2} \approx R$ or $\sigma_{x2} >> R$. In this case, the variable w_2 is dominated by its normally distributed components h_y^2 and h_x^2 and again can be well approximated as a normal variable.

Based on the normal approximation of w_1 and w_2, we can obtain from (3.25) and (3.27) that

$$\begin{aligned} p_z(z) = & \frac{1}{\sigma_{x1}}\phi_0\left(\frac{z}{\sigma_{x1}}\right)\Phi_0\left(\frac{-(R-z)^2-\tilde{D}(\sigma_{x2}^2-\sigma_{y2}^2)+(\sigma_{y1}^2+R^2)}{\sqrt{2\tilde{D}(\sigma_{y2}^4+\sigma_{x2}^4)+2\sigma_{y1}^2(\sigma_{y1}^2+2R^2)}}\right) \\ & +\frac{1}{\sigma_{y1}}\phi_0\left(\frac{z}{\sigma_{y1}}\right)\Phi_0\left(\frac{-(R-z)^2-\tilde{D}(\sigma_{y2}^2-\sigma_{x2}^2)+(\sigma_{x1}^2+R^2)}{\sqrt{2\tilde{D}(\sigma_{y2}^4+\sigma_{x2}^4)+2\sigma_{x1}^2(\sigma_{x1}^2+2R^2)}}\right) \end{aligned} \tag{3.28}$$

where the means and variances of w_1 and w_2 are calculated according to (3.10) and used in the arguments of $\Phi_0(\cdot)$.

From (3.28) and with integration by substitution, we can express the mean of z^2 as follows:

$$\begin{aligned} E(z^2) = & \frac{\sigma_{x1}^2}{\sqrt{2\pi}}\int_{-\infty}^{\infty} t^2 e^{-t^2/2}\Phi_0\left(\frac{-\sigma_{x1}^2t^2+2R\sigma_{x1}t-\tilde{D}(\sigma_{x2}^2-\sigma_{y2}^2)+\sigma_{y1}^2}{\sqrt{2\tilde{D}(\sigma_{y2}^4+\sigma_{x2}^4)+2\sigma_{y1}^2(\sigma_{y1}^2+2R^2)}}\right)dt \\ & +\frac{\sigma_{y1}^2}{\sqrt{2\pi}}\int_{-\infty}^{\infty} t^2 e^{-t^2/2}\Phi_0\left(\frac{-\sigma_{y1}^2t^2+2R\sigma_{y1}t-\tilde{D}(\sigma_{y2}^2-\sigma_{x2}^2)+\sigma_{x1}^2}{\sqrt{2\tilde{D}(\sigma_{y2}^4+\sigma_{x2}^4)+2\sigma_{x1}^2(\sigma_{x1}^2+2R^2)}}\right)dt \end{aligned} \tag{3.29}$$

As a special case, for the hyper-plane model (sphere model with R/σ_{x1} and $R/\sigma_{x2} \to \infty$, i.e., $z = \max\{x, y\}$), we have

$$E(z^2) = \frac{\sigma_{x1}^2}{\sqrt{2\pi}}\int_{-\infty}^{\infty} t^2 e^{-t^2/2}\Phi_0\left(\frac{\sigma_{x1}t}{\sigma_{y1}}\right)dt + \frac{\sigma_{y1}^2}{\sqrt{2\pi}}\int_{-\infty}^{\infty} t^2 e^{-t^2/2}\Phi_0\left(\frac{\sigma_{y1}t}{\sigma_{x1}}\right)dt = \frac{\sigma_{x1}^2+\sigma_{y1}^2}{2} \tag{3.30}$$

by using the equation

$$\int_{-\infty}^{\infty} t^2 e^{-t^2/2}\Phi_0(at)dx = \frac{\sqrt{2\pi}}{2}.$$

From (3.30) and (3.18), we have

$$\sigma_{z1} = \sqrt{\frac{\pi-1}{\pi}\frac{\sigma_{x1}^2+\sigma_{y1}^2}{2}} = \sqrt{\frac{\pi-1}{\pi}(1+kF^2)}\sigma_{x1}, \tag{3.31}$$

for the hyper-plane model.

3.4.4 *Calculation of σ_{z2}*

The variance σ_{z2}^2 of z_i ($i \neq 1$) is related to $E(h_z^2)$ by the following equation:

$$E(h_z^2) = E(z_2^2 + \cdots + z_D^2) = \tilde{D}E(z_2^2) = \tilde{D}\sigma_{z2}^2, \tag{3.32}$$

because all z_i's ($i \neq 1$) are i.i.d. with zero mean (Property 3.4).

According to (3.9), the pdf of h_z^2 is given by

$$\begin{aligned} p_{h_z^2}(w) &= p_{h_x^2}(w)\Pr\{(R-x)^2-(R-y)^2-h_y^2 < -w\} \\ &+ p_{h_y^2}(w)\Pr\{(R-y)^2-(R-x)^2-h_x^2 \le -w\}. \end{aligned} \tag{3.33}$$

Similar to the analysis of w_1 and w_2, both random variables $u_1 = (R-x)^2-(R-y)^2-h_y^2$ and $u_2 = (R-y)^2-(R-x)^2-h_x^2$ can be approximated as normally distributed if DE operates on the sphere model with $\sigma_{x2} \ge \sigma_{x1}$ or $\sigma_{x2} \approx \sigma_{x1}$. Using the approximate normal distribution of (3.27) and from (3.10), we can rewrite (3.33) as follows:

$$p_{h_z^2}(w) = \frac{1}{\sqrt{2\tilde{D}}\sigma_{x2}^2}\phi_0\left(\frac{w-\tilde{D}\sigma_{x2}^2}{\sqrt{2\tilde{D}}\sigma_{x2}^2}\right)\Phi_0\left(\frac{-w-(\sigma_{x1}^2-\sigma_{y1}^2-\tilde{D}\sigma_{y2}^2)}{\sqrt{2(\sigma_{x1}^4+\sigma_{y1}^4)+4R^2(\sigma_{x1}^2+\sigma_{y1}^2)+2\tilde{D}\sigma_{y2}^4}}\right)+(*),$$

where the second term $(*)$ is analogous to the first one by exchanging subscripts x and y.

By simple manipulations, the mean of h_z^2 is given by

$$\begin{aligned} E(h_z^2) &= \frac{1}{\sqrt{2\pi}}\int_{-\infty}^{\infty}(\tilde{D}\sigma_{x2}^2+\sqrt{2\tilde{D}}\sigma_{x2}^2 t)e^{-t^2/2} \\ &\cdot\Phi_0\left(\frac{-\sqrt{2\tilde{D}}\sigma_{x2}^2 t-\tilde{D}(\sigma_{x2}^2-\sigma_{y2}^2)-(\sigma_{x1}^2-\sigma_{y1}^2)}{\sqrt{2(\sigma_{x1}^4+\sigma_{y1}^4)+4R^2(\sigma_{x1}^2+\sigma_{y1}^2)+2\tilde{D}\sigma_{y2}^4}}\right)dt+(*) \end{aligned}$$

This expression can be simplified by using the equation (See the appendix in Section 3.7.2 for proof)

$$\begin{aligned} &\int_{-\infty}^{\infty}\frac{x}{\sigma_x}\phi_0\left(\frac{x-\mu_x}{\sigma_x}\right)\Phi_0\left(\frac{\mu_y-x}{\sigma_y}\right)dx \\ &= \mu_x\Phi_0\left(\frac{\mu_y-\mu_x}{\sqrt{\sigma_x^2+\sigma_y^2}}\right)-\frac{\sigma_x^2}{\sqrt{2\pi(\sigma_x^2+\sigma_y^2)}}\exp\left(-\frac{1}{2}\frac{(\mu_y-\mu_x)^2}{\sigma_x^2+\sigma_y^2}\right). \end{aligned} \tag{3.34}$$

Simple algebraic manipulation yields

$$E(h_z^2) = \tilde{D}\left[\sigma_{x2}^2\Phi_0\left(-\frac{\mu_-}{\sigma_+}\right)+\sigma_{y2}^2\Phi_0\left(\frac{\mu_-}{\sigma_+}\right)-\frac{2(\sigma_{x2}^4+\sigma_{y2}^4)}{\sqrt{2\pi}\sigma_+}\exp\left(-\frac{\mu_-^2}{2\sigma_+^2}\right)\right]. \tag{3.35}$$

where

$$\mu_- := \mu_{r_x^2}-\mu_{r_y^2} = \tilde{D}(\sigma_{x2}^2-\sigma_{y2}^2)+(\sigma_{x1}^2-\sigma_{y1}^2),$$

and

$$\sigma_+ := \sqrt{\sigma_{r_x^2}^2+\sigma_{r_y^2}^2} = \sqrt{4R^2(\sigma_{x1}^2+\sigma_{y1}^2)+2(\sigma_{x1}^4+\sigma_{y1}^4)+2\tilde{D}(\sigma_{x2}^4+\sigma_{y2}^4)}.$$

Thus, according to (3.32), we have

$$\sigma_{z2}^2 = \sigma_{x2}^2\Phi_0\left(-\frac{\mu_-}{\sigma_+}\right)+\sigma_{y2}^2\Phi_0\left(\frac{\mu_-}{\sigma_+}\right)-\frac{2(\sigma_{x2}^4+\sigma_{y2}^4)}{\sqrt{2\pi}\sigma_+}\left(-\frac{\mu_-^2}{2\sigma_+^2}\right). \tag{3.36}$$

3.4.5 From $\{\mathbf{z}_{i,g}\}$ to $\{\mathbf{x}_{i,g+1}\}$

In the last few subsections, we have analyzed the evolution of the mean and variance from $\{\mathbf{x}_{i,g}\}$, $\{\mathbf{y}_{i,g}\}$ to $\{\mathbf{z}_{i,g}\}$. The results apply to both the ADE and SDE models. We next consider the transformation from $\{\mathbf{z}_{i,g}\}$ to $\{\mathbf{x}_{i,g+1}\}$ based on the iterative mapping of ADE in (3.7). That is, we establish the relationship between the sample statistics of $\{\mathbf{z}_{i,g}\}$ involved in the right-hand sides of (3.7) and the mean $E(z)$ and variances σ_{z1}^2 and σ_{z2}^2 of $\mathbf{z}_{i,g}$ derived above.

According to Property 2, the pair $(\mathbf{x}_{i,g}, \mathbf{y}_{i,g})$ is identically distributed for different i, and so is $\mathbf{z}_{i,g}$ as a function (3.2) of $\mathbf{x}_{i,g}$ and $\mathbf{y}_{i,g}$. The pair $(\mathbf{x}_{i,g}, \mathbf{y}_{i,g})$ is generally not independent for different i due to the correlation between $\mathbf{y}_{i,g}$ and $\mathbf{x}_{j,g}$ $(i \neq j)$. This is because the indices r_0, r_1^i and r_2^i could be equal to j with probability $1/(NP-1)$. However, the correlation coefficient is at the order of $1/NP$ and approaches zero as the population size NP goes to infinity. Thus, when $\{\mathbf{x}_{i,g}\}$ is normally distributed (and so is $\{\mathbf{y}_{i,g}\}$), we can approximately assume that $(\mathbf{x}_{i,g}, \mathbf{y}_{i,g})$ are uncorrelated (i.e., independent) for different i. Furthermore, we approximately assume that the identically distributed $\mathbf{z}_{i,g}$'s are independent, and so are the elements of $\{z_{j,i,g}\}$.

According to statistical theory, both the sample mean and variance are unbiased statistics if the samples are i.i.d. Thus, we have

$$\begin{array}{ll} E(\bar{z}_{1,g}) = R_g - E(z_g), & E(s_{z1,g}^2) = \sigma_{z1,g}^2, \\ E(\bar{z}_{2,g}) = 0, & E(s_{z2,g}^2) = \sigma_{z2,g}^2. \end{array} \tag{3.37}$$

Then, we consider the average distance $E(r_{\bar{\mathbf{z}}})$ of the sample mean $\bar{\mathbf{z}}_g$ of $\{\mathbf{z}_{i,g}\}$ to the origin. Denote $\bar{\mathbf{z}}_g = \sum_{i=1}^{NP} \mathbf{z}_{i,g}/NP$ as $\bar{\mathbf{z}}_g := (\bar{z}_{1,g}, \bar{z}_{2,g}, \ldots, \bar{z}_{D,g})$, where $\bar{z}_{j,g}$ is the sample mean of $\{z_{j,i,g}\}$ averaged over all individuals i. The squared distance is given by $r_{\bar{\mathbf{z}}}^2 = \bar{z}_{1,g}^2 + \bar{z}_{2,g}^2 + \cdots + \bar{z}_{D,g}^2$. We have

$$\begin{aligned} E(r_{\bar{\mathbf{z}}}^2) &= E(\bar{z}_{1,g}^2) + \tilde{D}E(\bar{z}_{2,g}^2) \\ &= E(\bar{z}_{1,g})^2 + \tilde{D}E(\bar{z}_{2,g})^2 + Var(\bar{z}_{1,g}) + \tilde{D}Var(\bar{z}_{2,g}) \\ &= (R_g - E(z_g))^2 + \frac{Var(z_{1,i,g}) + \tilde{D}Var(z_{2,i,g})}{NP} \\ &= (R_g - E(z_g))^2 + \frac{\sigma_{z1,g}^2 + \tilde{D}\sigma_{z2,g}^2}{NP}. \end{aligned} \tag{3.38}$$

where the first line is obtained because $z_{j,i,g}$'s $(j \neq 1)$ are i.i.d, and so are the sample mean $\bar{z}_{j,g}$ of $\{z_{j,i,g}\}$. The variance of sample mean $\bar{z}_{j,g}$ is $Var(\bar{z}_{j,g}) = Var(z_{j,i,g})/NP$. This contributes to the third line of the above equation.

In spite of the simple expression of $E(r_{\bar{\mathbf{z}}}^2)$, the mean of $r_{\bar{\mathbf{z}}}$ is difficult to calculate. However, the variance of $r_{\bar{\mathbf{z}}}$ is expected to be small because it represents the distance of the sample mean of NP vectors to the origin. Indeed, if $\mathbf{z}$ is nearly normal distributed, it can be shown that the standard deviation of $r_{\bar{\mathbf{z}}}$ is about $1/NP$ of the mean of $r_{\bar{\mathbf{z}}}$. Thus, we have

$$E(r_{\bar{\mathbf{z}}})^2 = E(r_{\bar{\mathbf{z}}}^2) + Var(r_{\bar{\mathbf{z}}}) \approx E(r_{\bar{\mathbf{z}}}^2), \tag{3.39}$$

especially when NP is large.

Substituting (3.37), (3.38), (3.39) into (3.7), we have

$$R_{g+1}^2 = (R_g - E(z_g))^2 + \frac{\sigma_{z1,g}^2 + (D-1)\sigma_{z2,g}^2}{NP}, \tag{3.40}$$

$$\sigma_{x2,g+1}^2 = \sigma_{z2,g}^2, \tag{3.41}$$

$$\sigma_{x1,g+1}^2 = \sigma_{z1,g}^2. \tag{3.42}$$

These equations, together with the expressions of $E(z)$, σ_{z1} and σ_{z2} derived before, complete the analysis of DE from generation g to $g+1$. In addition, they form an iterative mapping which can be used to understand the complete evolutionary dynamics of DE for the sphere model.

3.5 Numerical Evaluation and Discussions

In this section, we compare the derived analytical results with practical DE experiments. As stated before, our analytical result is thought to be desirable only for optimization problem of dimension $D \geq 30$ and when the population size NP is rather large. We first conduct investigation when $D = 30$, $NP = 200$, and $F = 0.9$ or 0.5 (these are the two typical values used in the literature). Then, systematic experiments are conducted for a large range of D and NP to identify the applicability and the limitation of the analytical model.

3.5.1 *Performance Metrics*

The progress rate ϕ_R is a central quantity which has been used in the literature [11], [12] to characterize the effectiveness of an evolutionary algorithm. ϕ_R refers to the change of the residual distance of population mean to the optimum from generation g to $g+1$, i.e.,

$$\phi_{R,g} = R_g - R_{g+1}.$$

In view of its R-independent and D-dependent property, a normalized progress rate is introduced as follows:

$$\phi_{R,g}^* = \phi_{R,g}\frac{D}{R_g}.$$

The normalized variances of the parent population $\{\mathbf{x}_{i,g}\}$ are defined in a similar way:

$$\sigma_{x1,g}^* = \sigma_{x1,g}\frac{D}{R_g} \text{ and } \sigma_{x2,g}^* = \sigma_{x2,g}\frac{D}{R_g}.$$

Thus, the relative changes of σ_{x1}^* and σ_{x2}^* are given by

$$\phi_{\sigma 1,g}^* = \frac{\sigma_{x1,g+1}^*}{\sigma_{x1,g}^*} \text{ and } \phi_{\sigma 2,g}^* = \frac{\sigma_{x2,g+1}^*}{\sigma_{x2,g}^*}.$$

It is clear that at the steady state of DE, if it exists, the following necessary condition needs to be satisfied:

$$\phi^*_{\sigma 1,g} = \phi^*_{\sigma 2,g} = 1. \tag{3.43}$$

Given fixed parameters k, F, D and NP of DE, the normalized progress rate is completely determined by the normalized variances σ^*_{x1} and σ^*_{x2} and thus becomes a constant as the algorithm goes into the steady state. This leads to a constant ratio of

$$R_{g+1}/R_g = 1 - \phi^*_{R,g}/D. \tag{3.44}$$

Thus we have a linear convergence rate in the steady state.

In the following, we investigate the evolution of the progress rate and variances based on the formulae derived in the last section. They are verified by simulations of DE. In addition, we analyze the steady state of DE in which a linear convergence rate is achieved. The parameter setting of DE is then discussed based on these analyses.

3.5.2 Progress Rate

Illustrated in Figure 3.4-(a) is the normalized progress rate ϕ^*_R as a function of σ^*_{x1} and σ^*_{x2} in the case of $k = 1$, $F = 0.8$, $D = 30$, and $NP = 600$. The analytical result is consistent with the simulation. The ratio $\sigma^*_{x1}/\sigma^*_{x2}$ has great effect on the progress rate. As is shown, the ϕ^*_R curves move upward as the ratio increases. It seems that an enhanced variance σ^*_{x1} in the *progress direction* (the direction from the population mean to the optimum) is able to improve the progress of the algorithm. However, this needs further investigation since the practically achievable progress rate depends not only on the behavior of ϕ^*_R but also on that of $\phi^*_{\sigma 1}$ and $\phi^*_{\sigma 2}$.

The normalized progress rate climbs fast when σ^*_{x1} and σ^*_{x2} are small and then decreases slowly after reaching the optimum. This is mainly because large variances cause greater perturbation of the population mean, which in turn compromises the progress of population mean toward the optimum. This observation is inferred from the second term of (3.40). As it decreases, the progress rate might become negative for sufficiently large variances. However, this does not conflict with the elitist selection procedure of DE since the elitism merely guarantees the positive progress of the best individual (but not the mean) of the population.

3.5.3 Evolution of σ^*_{x1} and σ^*_{x2}

The evolutionary behavior of the normalized variances σ^*_{x1} and σ^*_{x2} is illustrated in Figure 3.4 (b) – (d). The y-axis represents the relative change, $\phi^*_{\sigma 1}$ or $\phi^*_{\sigma 2}$, of the variance from generation g to $g+1$. It is clear that the analytical and experimental results have a good match in most regions except the upper-right corner of the plots. This is the region where $\sigma_{x1} \approx R$ and $\sigma^*_{x1}/\sigma^*_{x2} \geq 4$ and thus the normal approximation of (3.27) in the derivation of σ_{z1} and σ_{z2} is inaccurate. The situation of $\sigma^*_{x1}/\sigma^*_{x2} \geq 4$, however, rarely happens during the evolutionary search of DE on the sphere model,

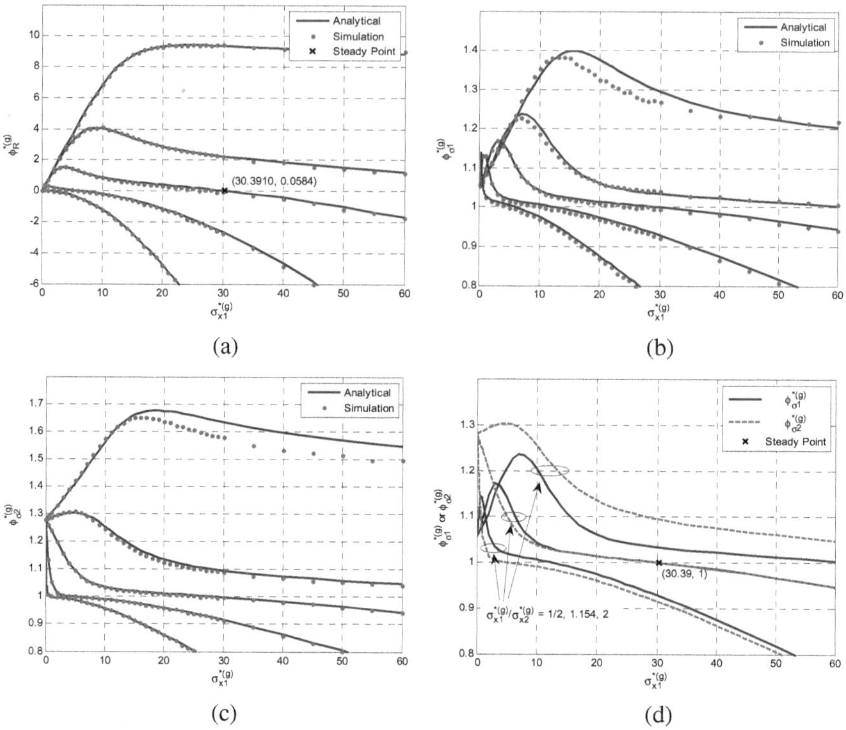

Fig. 3.4 The evolution of DE ($k = 1$, $F = 0.8$, $D = 30$, $NP = 600$). In descent order of their y axis values for $\sigma^*_{x1,g} \geq 30$, the five curves in plots (a) – (c) show the respective results for $\sigma^*_{x1,g}/\sigma^*_{x1,g}$ = 4, 2, 1.154, 1/2, and 1/4. The steady point in plot (d) is obtained as the solution to the equation $\phi^*_{\sigma 1}/\phi^*_{\sigma 2} = 1$. At the steady point, we have $\sigma^*_{x1,g} = 30.39$, $\sigma^*_{x1,g}/\sigma^*_{x1,g} = 1.154$ and $\phi^*_{R,g} = 0.05840$. All the simulation data points are obtained by averaging over 500 runs of one-generation experiments

because the selection pressure is generally greater in the progress direction than in other orthogonal directions.

As is shown from Figure 3.4 (b) and (c), both $\phi^*_{\sigma 1}$ and $\phi^*_{\sigma 2}$ curves are unimodal functions of σ^*_{x1} (given a fixed ratio $\sigma^*_{x1}/\sigma^*_{x2}$) and their asymptotic behavior is similar when σ^*_{x1} and σ^*_{x2} are large ($\sigma^*_{x1},\sigma^*_{x2} > D$, i.e., $\sigma_{x1}, \sigma_{x2} > R$). This similarity is more obvious if we draw the curves in the single plot (d). Let us restrict attention to the situation of $\sigma_{x1}, \sigma_{x2} > R$ where the parent population becomes roughly centered at the optimum point. If $\sigma^*_{x1}/\sigma^*_{x2}$ is much greater than 1, σ^*_{x1} is expected to decrease faster (or increase slower) than σ^*_{x2} due to the selection. Thus, the $\phi^*_{\sigma 1}$ curve is lower than that of $\phi^*_{\sigma 2}$. This is the case of the uppermost pair of curves in the plot. Similarly, in the opposite case when $\sigma^*_{x1}/\sigma^*_{x2}$ is much less than 1, it is the curve $\phi^*_{\sigma 2}$ that is lower than the other one. Between the above situations, there is a special case of $\sigma^*_{x1}/\sigma^*_{x2} \approx 1$[1]

[1] The $\phi^*_{\sigma 2}$ and $\phi^*_{\sigma 2}$ curves intersect with $\sigma^*_{x1}/\sigma^*_{x2} = 1$ if the parent population exactly centers at the optimum point (i.e., $\sigma^*_{x1}, \sigma^*_{x2} \to \infty$).

in which the two curves overlap for large σ_{x1}^* and σ_{x2}^*. In plot (d), analytical results indicate a ratio of $\sigma_{x1}^*/\sigma_{x2}^* = 1.154$. This value is slightly overestimated. In fact, in the region where the analytical curves overlap, the corresponding simulation results in plots (b) and (c) show that the $\phi_{\sigma 1}^*$ curve is slightly lower than the $\phi_{\sigma 2}^*$ curve. The curve overlap of simulation results happens with a $\sigma_{x1}^*/\sigma_{x2}^*$ closer to 1.

3.5.4 Steady Point

It is meaningful to consider the steady state of DE where the $\phi_{\sigma 1}^*$ and $\phi_{\sigma 2}^*$ curves intersect with $\phi_{\sigma 1}^* = \phi_{\sigma 2}^* = 1$. As shown in Figure 3.4-(d), the solution to $\phi_{\sigma 1}^* = \phi_{\sigma 2}^* = 1$ is that $\sigma_{x1}^* = 30.39$ and $\sigma_{x1}^*/\sigma_{x2}^* = 1.154$ (i.e., $\sigma_{x2}^* = 26.33$). Both $\phi_{\sigma 1}^*$ and $\phi_{\sigma 2}^*$ are decreasing functions in the neighborhood of $\sigma_{x1}^* = 30.39$. If $\sigma_{x1}^* < 30.39$, then both $\phi_{\sigma 1}^*$ and $\phi_{\sigma 2}^*$ are less than 1 and the values of σ_{x1}^* and σ_{x2}^* decrease at generation g+1. In the opposite case $\sigma_{x1}^* > 30.39$, $\phi_{\sigma 1}^*$ and $\phi_{\sigma 1}^*$ are greater than 1 and σ_{x1}^* and σ_{x2}^* increase. This implies that σ_{x1}^* and σ_{x2}^* tend to remain in the neighborhood of $(\sigma_{x1}^*, \sigma_{x2}^*)$= (30.39, 26.33). It is conjectured that these pair values correspond to the steady state of DE. The stability is verified by simulations, although a strict proof is difficult to provide.

Corresponding to the steady point in plot (d), we can obtain from plot (a) that $\phi_R^* = 0.05840$. It corresponds to a linear convergence rate of $R_{g+1}/R_g = 0.9981$ according to (3.44). The rate is much smaller than the maximum achievable value as indicated in plot (a). This is mainly because the variances σ_{x1}^* and σ_{x2}^* are too large at the steady state. The progress rate can be improved by decreasing the mutation factor F; e.g., a normalized progress rate $\phi_R^* = 0.4844$ can be achieved if $F = 0.5$ (the plots are similar to Figure 3.4 and not presented due to space limitation). However, even in this case, the progress rate is much less than the maximum because of the large variances σ_{x1}^* and σ_{x2}^*. This reminds us to be careful about the attempt to improve DE by merely increasing the variance σ_{x1}^* in the progress direction. A relatively small σ_{x1}^* and a larger $\sigma_{x1}^*/\sigma_{x2}^*$ seem more beneficial.

3.5.5 Dynamic Behavior of DE

Having understood the behavior from generation g to $g+1$, we next investigate the complete evolutionary dynamic of DE. With the same parameters of $k = 1$, $F = 0.8$, $D = 30$, and $NP = 600$, a complete record of the DE evolution dynamic is depicted over the period of 1,000 generations in Figure 3.5. The initial population is generated according to an isotropic normal distribution with mean (10^3, 0, ..., 0) and variance 1, i.e., $R^{(0)} = 10^3$ and $\sigma_{x1}^{(0)} = \sigma_{x2}^{(0)} = 1$. The analytical results are obtained by using the iterative mapping of (3.40) – (3.42), while the simulation results are averaged over 10 runs of DE experiments. The analytical results well predict the basic trends of simulation progress in spite of a major discrepancy: simulation results show a slower response to the poorly selected initial variances. This might be because the parent population becomes not as close to being normal distributed when the variances increase quickly. The exact reason, however, is not well understood.

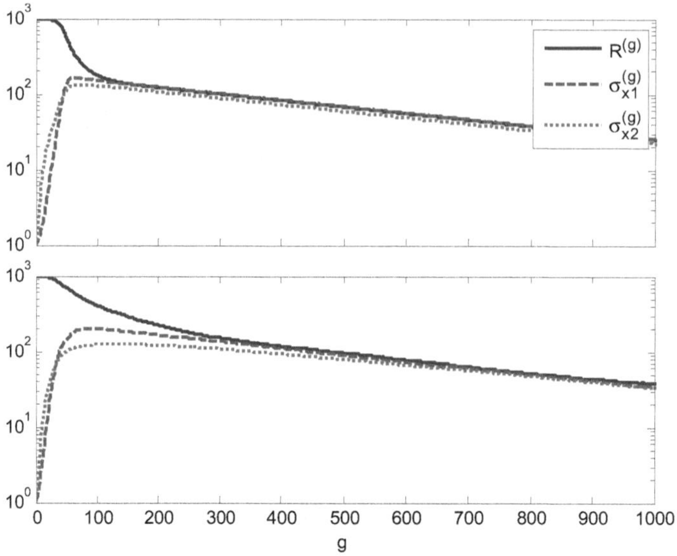

Fig. 3.5 The evolution dynamics of DE with $R_0 = 10^3$ and $\sigma_{x1,0} = \sigma_{x2,0} = 1$ ($k = 1, F = 0.8$, $D = 30$, $NP = 600$). Upper plot: analytical results. Lower plot: simulation results

3.5.6 *Effect of Mutation Factor*

Mutation factor F affects the exploration and exploitation capabilities of an algorithm. It also indirectly affects the progress rate by changing σ_{y1} and σ_{y2}. Valuable information can be observed from a comparison between the analytical results for F = 0.5 (Figure 3.6) and those for F = 0.8 (Figure 3.4 or Figure 3.5). As explained below, it is consistent with the general observations from experimental studies on the sphere model [72]: a relatively small mutation factor increases the convergence rate but may lead to more premature convergence.

As indicated in Figure 3.4-(a) and Figure 3.6-(a), the small mutation factor F = 0.5 leads to a larger achievable progress rate ϕ_R^*. However, Figure 3.6-(b) shows that the small F also leads to a small $\phi_{\sigma 1}^*$ which could be even less than 1 (i.e., $\sigma_{x1,g+1}^* < \sigma_{x1,g}^*$) if the variance $\sigma_{x1,g}^*$ is small. This is an implication of less exploration capability and explains the evolution dynamics in Figure 3.6-(c) where the algorithm is initialized with small $\sigma_{x1,0}^* = \sigma_{x1,0}D/R_0 = 0.03$.

At the first stage of Figure 3.6-(c), σ_{x1} decreases ($\phi_{\sigma 1}^* < 1$) and σ_{x2} increases ($\phi_{\sigma 2}^* > 1$); thus the ratio $\sigma_{x1,0}^*/\sigma_{x2,0} = \sigma_{x1}/\sigma_{x2}$ becomes smaller and both $\phi_{\sigma 1}^*$ and $\phi_{\sigma 2}^*$ get closer to 1, as implied by Figure 3.6-(b). If $\phi_{\sigma 1}^*$ becomes equal to 1 earlier than $\phi_{\sigma 2}^*$ does (i.e., both $\phi_{\sigma 1}^*$ and $\phi_{\sigma 2}^*$ are greater than 1), then both σ_{x1} andσ_{x2} increase until they are close to R and the algorithm goes into the steady state. This is the case of F = 0.5 as is shown in Figure 3.6 (c). If instead $\phi_{\sigma 2}^*$ becomes equal to 1 first (i.e., both $\phi_{\sigma 1}^*$ and $\phi_{\sigma 2}^*$ are less than 1), then both σ_{x1} andσ_{x2} decrease simultaneously and approaches zero very fast, leading to a premature convergence. This happens if the mutation factor F is below a certain lower limit.

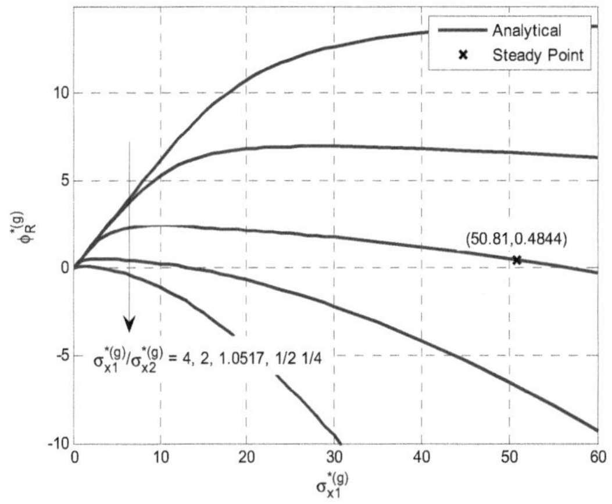

(a)

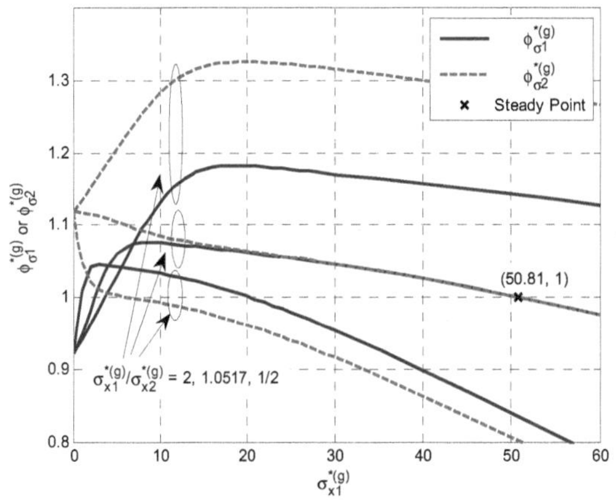

(b)

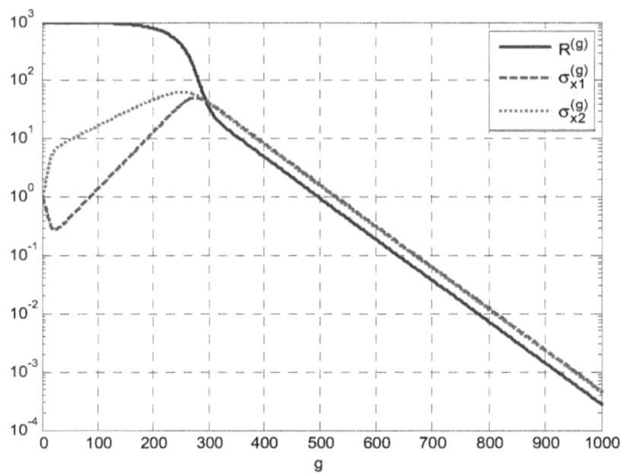

(c)

Fig. 3.6 The evolution of DE (k = 1, F = 0.5, D = 30, NP = 600). (a): ϕ_R^*, (b): $\phi_{\sigma1}^*$ and $\phi_{\sigma2}^*$, and (c): evolution dynamics with $R_0 = 10^3$ and $\sigma_{x1,0} = \sigma_{x1,0}$= 1

As we have investigated, if the algorithm is initialized with a ratio $\sigma_{x1,0}/R_0 = \sigma_{x2,0}/R_0 = 10^{-3}$, the smallest mutation factor that avoids premature convergence is equal to 0.44. If the ratio decreases to 10^{-6}, a higher limit of $F > 0.48$ is necessary. In the extreme case of $\sigma_{x1,0}/R_0 = \sigma_{x2,0}/R_0 \to 0$ (i.e., the hyper-sphere landscape becomes a hyper-plane), σ_{x2} has no effect on σ_{x1} any more and we can simply use equation (3.31) to determine the smallest value of F. We have

$$\sqrt{\frac{\pi-1}{\pi}(1+kF^2)} > 1, \tag{3.45}$$

i.e., $F > 0.68$ in the case of $k = 1$. If $F \leq 0.68$, the variance σ_{x1} tends to decrease as σ_{x2} goes up exponentially. During this period, there is little evolution progress in the direction towards the optimum because $\sigma_{x1} \to 0$.

From the above analysis, it is clear that there is a balance between the speed and the reliability of the algorithm's convergence. The former prefers a relatively small mutation factor, while the latter favors a relatively large value. There is no single mutation factor that fulfills both objectives. The algorithm that best balances the two objectives should be the one that is capable of varying the mutation factor in an adaptive manner as the algorithm proceeds. For the sphere model, the mutation factor can be initialized to be as high as 0.68 in case the initial population is poorly located far from the optimum. As the population approaches the optimum, the lower limit of F to avoid premature convergence decreases and thus the mutation factor can be accordingly set to a slightly smaller value to improve the convergence rate. The problem, however, is how to design a simple yet effective scheme to adapt the mutation factor in different evolution stages.

3.6 Summary

In this chapter, we have proposed an analytical method to investigate the evolutionary stochastic properties of DE on the sphere model. Resorting to the geometric symmetry of the sphere model, we have identified the property of rotational invariance which is satisfied by the parental and intermediate populations at each generation. This property is utilized to introduce an approximate model of DE, based on which the analysis of the evolutionary dynamics of DE can be done in a mathematical manner.

The derived formulae are verified by simulation and used to provide some insights into the parameter setting of DE. It is shown that a mutation factor as high as 0.68 is required to avoid premature convergence for the sphere model if the initial population is located remotely from the optimum (i.e., the spherical landscape becomes close to a hyper-plane). When the population gets closer to the optimum, a relatively small mutation factor becomes beneficial for the sake of faster convergence. It clearly shows that there is no fixed mutation factor that is capable of adaptively balancing the speed and the reliability of convergence at difference evolution stages even for the simple sphere model. This encourages efforts to improve

DE by dynamically adapting the control parameters to appropriate values during the optimization process.

3.7 Appendix

3.7.1 Proof of Property 3.3

It is equivalent to prove that if $p^g_{\mathbf{x}}(\mathbf{x})$ is rotationally symmetric around the first coordinate axis, so are $p^g_{\mathbf{y}}(\mathbf{y})$, $p^g_{\mathbf{z}}(\mathbf{z})$ and $p^g_{\mathbf{x}}(\mathbf{x})$. The rotational symmetry of $p^g_{\mathbf{x}}(\mathbf{x})$ can be expressed as $p^g_{\mathbf{x}}(\mathbf{x}) = p^g_{\mathbf{x}}(\mathbf{x}')$, for $\forall \mathbf{x} = (x_1, x_2, \ldots, x_D) =: (x_1, \mathbf{x}_{2:D})$ and $\mathbf{x}' = (x_1, x'_2, \ldots, x'_D) =: (x_1, \mathbf{x}'_{2:D})$ where $\sum_{i=2}^D x_i^2 = \sum_{i=2}^D x'^2_i$ or $\mathbf{x}'_{2:D} = \mathbf{x}_{2:D}U$ for a certain rotation (i.e., orthonormal) matrix U.

First consider the distribution of $\{\mathbf{y}_{i,g}\}$. According to (3.1), $\mathbf{y}_{i,g}$ is a linear combination of i.i.d. random vectors $\mathbf{x}_{r_0,g}$, $\mathbf{x}_{r^i_1,g}$ and $\mathbf{x}_{r^i_2,g}$. Consider any arbitrary variable $\tilde{\mathbf{y}} = c_1\mathbf{x}_1 + c_2\mathbf{x}_2$ which is a linear combination of two i.i.d. variables $\mathbf{x}_a$ and $\mathbf{x}_b$ following the distribution $p^g_{\mathbf{x}}(\mathbf{x})$. The pdf of $\tilde{\mathbf{y}}$ is the convolution of the pdfs of $c_1\mathbf{x}_1$ and $c_2\mathbf{x}_2$, i.e., $p^g_{\mathbf{x}}(\mathbf{x}/c_1)/c_1$ and $p^g_{\mathbf{x}}(\mathbf{x}/c_2)/c_2$. We have

$$p^g_{\tilde{\mathbf{y}}}(\tilde{\mathbf{y}}) = \int_{\mathbf{x}} \frac{1}{c_1} p^g_{\mathbf{x}}\left(\frac{\mathbf{x}}{c_1}\right) \frac{1}{c_2} p^g_{\mathbf{x}}\left(\tilde{\mathbf{y}} - \frac{\mathbf{x}}{c_2}\right) dx. \tag{3.46}$$

Consider any two vectors $\tilde{\mathbf{y}} = (\tilde{y}_1, \tilde{y}_2, ..., \tilde{y}_D) =: (\tilde{y}_1, \tilde{\mathbf{y}}_{2:D})$ and $\tilde{\mathbf{y}}' = (\tilde{y}_1, \tilde{y}'_2, ..., \tilde{y}'_D) =: (\tilde{y}_1, \tilde{\mathbf{y}}'_{2:D})$ where $\tilde{\mathbf{y}}'_{2:D} = \tilde{\mathbf{y}}_{2:D}U$ for a certain orthonormal matrix U. We have

$$\begin{aligned}
p^g_{\tilde{\mathbf{y}}}(\tilde{\mathbf{y}}') &= \int_{\mathbf{x}'} \tfrac{1}{c_1} p^g_{\mathbf{x}}\left(\tfrac{\mathbf{x}'}{c_1}\right) \tfrac{1}{c_2} p^g_{\mathbf{x}}\left(\tilde{\mathbf{y}}' - \tfrac{\mathbf{x}'}{c_2}\right) d\mathbf{x}' \\
&= \int_{\mathbf{x}'} \tfrac{1}{c_1} p^g_{\mathbf{x}}\left(\tfrac{x_1}{c_1}, \tfrac{\mathbf{x}'_{2:D}}{c_1}\right) \tfrac{1}{c_2} p^g_{\mathbf{x}}\left(\tilde{y}_1 - \tfrac{x_1}{c_2}, \tilde{\mathbf{y}}'_{2:D} - \tfrac{\mathbf{x}'_{2:D}}{c_2}\right) dx_1 d\mathbf{x}'_{2:D} \\
&= \int_{\mathbf{x}} \tfrac{1}{c_1} p^g_{\mathbf{x}}\left(\tfrac{x_1}{c_1}, \tfrac{\mathbf{x}_{2:D}U}{c_1}\right) \tfrac{1}{c_2} p^g_{\mathbf{x}}\left(\tilde{y}_1 - \tfrac{x_1}{c_2}, \tilde{\mathbf{y}}_{2:D}U - \tfrac{\mathbf{x}_{2:D}U}{c_2}\right) dx_1 d\mathbf{x}_{2:D} \\
&= \int_{\mathbf{x}} \tfrac{1}{c_1} p^g_{\mathbf{x}}\left(\tfrac{x_1}{c_1}, \tfrac{\mathbf{x}_{2:D}}{c_1}\right) \tfrac{1}{c_2} p^g_{\mathbf{x}}\left(\tilde{y}_1 - \tfrac{x_1}{c_2}, \tilde{\mathbf{y}}_{2:D} - \tfrac{\mathbf{x}_{2:D}}{c_2}\right) dx_1 d\mathbf{x}_{2:D}
\end{aligned}$$

where the first line is obtained from (3.46) by using a dummy variable $\mathbf{x}'$ instead of $\mathbf{x}$. The third line follows from integration by substitution $\mathbf{x}'_{2:D} = \mathbf{x}_{2:D}U$ and is obtained by noting that $d\mathbf{x}'_{2:D} = d\mathbf{x}_{2:D}$ because the determinant of any orthonormal matrix $U$is equal to 1. The last line is due to the rotational-symmetry property of $p^g_{\mathbf{x}}(\mathbf{x})$. Since the last line has the same expression as in (3.46), we have proven the rotational symmetry of $\tilde{y}$. As a result, we can repeat the above proof and show the rotational symmetry for any linear combination of multiple independently isotropically distributed variables. As a special case, the rotational symmetry of $\mathbf{y}_{i,g}$ can be established.

We next consider the distribution of $\{\mathbf{z}_{i,g}\}$. According to (3.2), the probability distribution $p^g_{\mathbf{z}}(\mathbf{z})$ is given as follows

$$
\begin{aligned}
p_{\mathbf{z}}^{g}(\mathbf{z}) &= p_{\mathbf{x}}^{g}(\mathbf{z})\Pr\{r_{\mathbf{x}} \le r_{\mathbf{y}}|\mathbf{x}=\mathbf{z}\} + p_{\mathbf{y}}^{g}(\mathbf{z})\Pr\{r_{\mathbf{x}} > r_{\mathbf{y}}|\mathbf{y}=\mathbf{z}\} \\
&= p_{\mathbf{x}}^{g}(\mathbf{z})\Pr\{r_{\mathbf{y}} \ge r_{\mathbf{z}}\} + p_{\mathbf{y}}^{g}(\mathbf{z})\Pr\{r_{\mathbf{x}} > r_{\mathbf{z}}\} \\
&= p_{\mathbf{x}}^{g}(\mathbf{z})\int_{\|\mathbf{y}\|\ge\|\mathbf{z}\|} p_{\mathbf{y}}^{g}(\mathbf{y})d\mathbf{y} + p_{\mathbf{y}}^{g}(\mathbf{z})\int_{\|\mathbf{x}\|>\|\mathbf{z}\|} p_{\mathbf{x}}^{g}(\mathbf{x})d\mathbf{x}.
\end{aligned}
\tag{3.47}
$$

Denote $\mathbf{z} = (z_1, z_2, \ldots, z_D)$ and $\mathbf{z}' = (z_1, z_2', \ldots, z_D')$ where $\sum_{i=2}^{D} z'^2_i = \sum_{i=2}^{D} z_i^2$. We have $p_{\mathbf{x}}^{g}(\mathbf{z}) = p_{\mathbf{x}}^{g}(\mathbf{z}')$, $p_{\mathbf{y}}^{g}(\mathbf{z}) = p_{\mathbf{y}}^{g}(\mathbf{z}')$ and $\| \mathbf{z} \| = \| \mathbf{z}' \|$. Thus, we have $p_{\mathbf{z}}^{g}(\mathbf{z}) = p_{\mathbf{z}}^{g}(\mathbf{z}')$ according to the last line of (3.47). Thus, $p_{\mathbf{z}}^{g}(\mathbf{z})$ is rotationally symmetric around the first coordinate axis. This also proves the rotational symmetry of $p_{\mathbf{x}}^{g}(\mathbf{x})$ since $\mathbf{x}_{i,g+1} = \mathbf{z}_{i,g}$ in DE.

The property of rotational symmetry can be proved for the sample means $\bar{\mathbf{y}}_g$, $\bar{\mathbf{z}}_g$ and $\bar{\mathbf{x}}_{g+1}$ in a similar, standard manner. The proof is omitted due to space limitation.

3.7.2 Proof of Eqs. (3.16) and (3.34)

By substituting $t = (x - \mu_x)/\sigma_x$ and noting that $\phi_0(t) = 1/\sqrt{2\pi}exp(-t^2/2)$, we have

$$
\begin{aligned}
&\int_{-\infty}^{\infty} \frac{x}{\sigma_x}\phi_0\left(\frac{x-\mu_x}{\sigma_x}\right)\Phi_0\left(\frac{x-\mu_y}{\sigma_y}\right)dx \\
&= \frac{1}{\sqrt{2\pi}}\int_{-\infty}^{\infty}(\sigma_x t + \mu_x)e^{-t^2/2}\Phi_0\left(\frac{\sigma_x t+\mu_x-\mu_y}{\sigma_y}\right)dt
\end{aligned}
$$

This expression can be simplified by using the equations [12]

$$\int_{-\infty}^{\infty} e^{-t^2/2}\Phi_0(at+b)dx = \sqrt{2\pi}\Phi_0\left(\frac{b}{\sqrt{1+a^2}}\right)$$

and

$$\int_{-\infty}^{\infty} te^{-t^2/2}\Phi_0(at+b)dx = \frac{a}{\sqrt{1+a^2}}\exp\left(-\frac{1}{2}\frac{b^2}{1+a^2}\right)$$

Thus, by simple manipulation, we have

$$
\begin{aligned}
\int_{-\infty}^{\infty} \frac{x}{\sigma_x}\phi_0\left(\frac{x-\mu_x}{\sigma_x}\right)\Phi_0\left(\frac{x-\mu_y}{\sigma_y}\right)dx &= \mu_x\Phi_0\left(\frac{\mu_y-\mu_x}{\sqrt{\sigma_x^2+\sigma_y^2}}\right) \\
&\quad + \frac{|\sigma_y|}{\sigma_y}\frac{\sigma_x^2}{\sqrt{2\pi(\sigma_x^2+\sigma_y^2)}}\exp\left(-\frac{1}{2}\frac{(\mu_y-\mu_x)^2}{\sigma_x^2+\sigma_y^2}\right).
\end{aligned}
$$

Equations (3.16) and (3.34) are merely special cases of the above equation.

Chapter 4
Parameter Adaptive Differential Evolution

The performance of differential evolution is affected by its control parameters which in turn are dependent on the landscape characteristics of the objective function. As is clear from extensive experimental studies and theoretical analysis of simple spherical functions, inappropriate control parameter settings may lead to false or slow convergence and therefore degrade the optimization performance of the algorithm. To address this problem, one method is to automatically update control parameters based on the feedback from the evolutionary search. As better parameter values tend to generate offspring that are more likely to survive, parameter adaptation schemes usually follow the principle of propagating parameter values that have produced successful offspring solutions.

In this chapter, a new differential evolution algorithm, JADE, is proposed to update control parameters in such an adaptive manner that follows the principle described above. Different from other adaptive methods which are usually based on the classic DE/rand/1/bin, JADE implements two new greedy mutation strategies to make use of the direction information provided by both the high-quality solutions in the current population and the archived inferior solutions recently explored in the evolutionary search.

Simulation results show that the proposed parameter adaptive differential evolution, JADE, is capable of evolving control parameters to appropriate values for different problems and at different evolution stages of a single problem. It also indicates that the proposed parameter adaptation scheme and greedy mutation strategy are mutually beneficial. JADE is better than, or at least comparable to, many other competitive evolutionary algorithms from the literature in terms of convergence rate and reliability for a set of 20 benchmark problems It also shows overall better scalability for high dimensional problems.

4.1 Introduction

Differential evolution has been shown to be a simple yet efficient evolutionary algorithm for the optimization for many real-world problems [1], [2] when its control

J. Zhang and A.C. Sanderson: Adaptive Differential Evolution, ALO 1, pp. 39–82.
springerlink.com

parameters are set appropriately. To obtain the appropriate control parameter setting, a straightforward method is to tune the parameters by trial and error, which however requires tedious optimization trials even for an algorithm (e.g., the classic DE/rand/1/bin) which fixes the parameters throughout the evolutionary search. It is even harder, if not impossible, to use trial and error to find appropriate parameter values at different evolution stages.

To address the problem of tedious parameter tuning, different adaptive or self-adaptive mechanisms, [16] – [25] have been recently introduced to dynamically update the control parameters without the user's prior knowledge of the problem or interaction during the search process. As discussed in Chapter 2, adaptive or self-adaptive parameter control, if well designed, may enhance the robustness of an algorithm by automatically adapting the parameters to the characteristic of different fitness landscapes and thus being applicable to various optimization problems without trial and error. Furthermore, the convergence rate can be improved when the control parameters are adapted to appropriate values at different evolution stages of an optimization problem.

The adaptive or self-adaptive DE algorithms have shown faster and more reliable convergence performance than the classic DE algorithms without parameter control for many test problems [21] – [25]. It is of interest to find the most suitable DE variant and use it as the underlying scheme where adaptive operations can be introduced. Some adaptive algorithms [17], [22] are developed based on the classic DE/rand/1/bin, which is known to be robust but less efficient in terms of convergence rate. Others [21], [23], [24] simultaneously implement DE/rand/1/bin and one or more greedy DE variants such as DE/current-to-best and dynamically adapt the probability of using each variant to generate offspring. In these algorithms, the same parameter adaptation scheme is applied to tune the control parameters of different DE variants.

There has been no method developed solely based on a greedy DE variant (such as DE/current-to-best and DE/best) that utilizes the information of the best-so-far solutions in the current population. The reason seems straightforward: a greedy variant is usually less reliable and may lead to problems such as premature convergences especially when solving multi-modal problems [77], [9]. However, we note that a well-designed adaptation algorithm is usually very beneficial to enhance the robustness of an algorithm. Hence, for an adaptive DE algorithm, the reliability of greedy DE variants becomes a less crucial problem, while their fast convergence property becomes more attractive.

Other than the best solutions in the current population, the history data is another source that can be used to improve the convergence performance. One example is the particle swarm optimization (PSO) [78] where the best solution previously explored is used to direct the movement of the current population (swarm). In this chapter, instead of the *best* solution ever explored, we are interested in the *inferior* solutions in recently past generations and consider their difference with the current population as a promising progress direction towards the optimum. These solutions are also helpful to increase the population diversity.

In view of the above considerations, we introduce two new greedy DE mutation strategies, archived assisted 'DE/current-to-$p$best' and 'DE/rand-to-$p$best', and use either of them as the basis of our parameter adaptation scheme. The new mutation strategies utilize the direction information provided not only by the best solutions but also by the inferior solutions recently explored in the evolutionary search. The parameter adaptation scheme follows the basic principle of propagating parameter values that tends to generate successful offspring, and addresses the problem of the inconsistency between success probability and progress rate. The resultant algorithm, JADE, is therefore capable of evolving parameters to a value that tend to produce successful offspring with significant progress.

4.2 Adaptive DE Algorithms

In this section, we provide a brief review of recent adaptive and self-adaptive DE algorithms which dynamically update control parameters as the evolutionary search proceeds. It provides a convenient way to compare these adaptive DE algorithms with the JADE algorithm proposed later.

4.2.1 DESAP

Teo [17] proposed an algorithm, named DESAP, which not only dynamically adapts mutation and crossover parameters η and δ (which is usually the case in other adaptive or self-adaptive algorithms) but also the population size π. The base strategy of DESAP is slightly different from the classic DE/rand/1/bin and rather similar to the scheme introduced in [16]: while δ and π have the same meanings as CR and NP, respectively, η refers to the probability of applying an additional normally-distributed mutation after crossover. The ordinary mutation factor F is indeed kept fixed in DESAP.

In DESAP, each individual i is associated with its control parameters δ_i, η_i and π_i. The adaptation of these parameters is realized by evolving them through crossover and mutation, just as the operations applied to each individual $\mathbf{x}_i$. The new values of these control parameters survive together with $\mathbf{u}_i$ if $f(\mathbf{u}_i) < f(\mathbf{x}_i)$. In spite of its simple reasoning, DESAP's performance is not satisfactory. It outperforms the conventional DE only in one of the five De Jong's test problems, while other results are very similar. Indeed, as the author states, DESAP is mainly used to demonstrate a possibility of further reducing control parameters by adaptively updating the population size like other parameters.

4.2.2 FADE

The fuzzy adaptive differential evolution (FADE), introduced by Liu and Lampinen [19], is a new variant of DE, which uses fuzzy logic controllers to adapt the control parameters F_i and CR_i for the mutation and crossover operations. A similar method has also been independently proposed for multi-objective DE optimization [20]. Similar to many other adaptive DE algorithms (except DESAP), the population size is assumed to be tuned in advance and kept fixed throughout the evolution process of FADE. The fuzzy-logic controlled approach is tested with a set of 10 standard benchmark functions and shows better results than the classic DE when the problem's dimension is high.

4.2.3 SaDE

SaDE (Self-adaptive Differential Evolution) [21] is proposed by Qin and Suganthan to simultaneously implement two mutation strategies 'DE/rand/1' (strategy 1) and 'DE/current-to-best/2' (strategy 2). The parameter adaptation is composed of two parts: (i) the adaptation of the probability, p_i (i = 1 and 2), of applying the i-th mutation strategy; and (ii) the adaptation of the original two control parameters F_i and CR_i of DE.

The probability of generating a mutation vector according to either of the two strategies is initialized to be 0.5 and then updated every 50 generations in the following manner:

$$p_1 = \frac{ns_1(ns_2+nf_2)}{ns_2(ns_1+nf_1)+ns_1(ns_2+nf_2)}, p_2 = 1-p_1, \tag{4.1}$$

where ns_i and nf_i are the respective numbers of the offspring vectors generated by the i-th (i = 1, 2) strategy that survive or fail in the selection operation in the last 50 generations. It is believed that this adaptation procedure can gradually evolve the most suitable mutation strategy at different learning stages for the problem under consideration This is similar to the scheme proposed in [79], [80], where competing heuristics (including several DE variants, simplex methods and evolutionary strategies) are simultaneously adopted and the probabilities of generating offspring by any of these heuristics are adapted dynamically.

The mutation factors F_i are independently generated at each generation according to a normal distribution with mean 0.5, standard deviation 0.3,

$$F_i = \text{randn}(0.5, 0.3), \tag{4.2}$$

and truncated to the interval (0, 2]. It states that this scheme can keep both local (with small F values) and global (with large F values) search ability to generate potentially good mutant vectors throughout the evolution process.

The crossover probabilities CR_i are independently generated according to a normal distribution with mean CR_m and standard deviation 0.1, i.e.,

$$CR_i = \text{randn}(CR_m, 0.1). \tag{4.3}$$

To the contrary of F_i, the CR_i values generated above remain fixed for 5 generations before the next re-generation. The mean CR_m is initialized to be 0.5. It is then updated every 25 generations and set to be the mean of successful CR values in the last 25 generations, i.e.,

$$CR_m = \frac{1}{K}\sum_{k=1}^{K} CR_{\text{suc}}(k), \tag{4.4}$$

where K is the number of these successful CR values and $CR_{\text{suc}}(k)$ denotes k-th value.

To speed up the convergence of SaDE, the authors further apply a local search procedure (Quasi–Newton method) to some good individuals after 200 generations.

While the original method is proposed to solve unconstrained optimization problems, SaDE is further extended by implementing five candidate mutation strategies in solving a set of constrained optimization problems [81].

4.2.4 SaNSDE

SaNSDE [25] is proposed by Yang et al. as an adaptive differential evolution algorithm which generates mutation vectors in the same method as SaDE except that the mutation factors in the two mutation strategies are generated according to either a normal or a Cauchy distribution , i.e.,

$$F_i = \begin{cases} \text{randn}(0.5, 0.3), & \text{if } rand < fp \\ \text{randc}(0.0, 1.0), & \text{otherwise,} \end{cases} \tag{4.5}$$

where randn(μ, σ^2) denotes a random value from a normal distribution of mean μ and variance σ^2, and randc(μ, δ) a random value from a Cauchy distribution with location and scale parameters μ and δ, respectively. The probability fp of applying either of the two distributions in (4.5) is adapted in a manner similar to (4.1) in SaDE.

The adaptation of the crossover rate also follows the method in SaDE except that CR_m is the updated as a weighted average of the successful CR values,

$$CR_m = \frac{1}{K}\sum_{k=1}^{K} w_k CR_{\text{suc}}(k), \tag{4.6}$$

every 25 generations, where the weight

$$w_k = \Delta_k \Big/ \sum_{k=1}^{K} \Delta_k \tag{4.7}$$

is calculated by normalizing the corresponding positive improvement $\Delta = f(\mathbf{x}) - f(\mathbf{u})$ in the selection, related to each successful crossover rate $CR_{\text{suc}}(k)$.

4.2.5 jDE

Brest et al. [22] proposed a new adaptive DE, jDE, which is based on the classic DE/rand/1/bin. Similar to other schemes, jDE fixes the population size during the optimization while adapting the control parameters F_i and CR_i associates with each individual. The initialization process sets $F_i = 0.5$ and $CR_i = 0.9$ for each individual. Instead of SaDE's method which simply randomizes F_i and updates CR_m by accumulated successful records, jDE regenerates (with probabilities $\tau_1 = \tau_2 = 0.1$ at each generation) new values for F_i and CR_i according to uniform distributions on [0.1, 1] and [0, 1], respectively. That is,

$$F_{i,g+1} = \begin{cases} 0.1 + 0.9\text{rand}_1, & \text{if rand}_2 < \tau_1 \\ F_{i,g}, & \text{otherwise} \end{cases} \tag{4.8}$$

and

$$CR_{i,g+1} = \begin{cases} \text{rand}_3, & \text{if rand}_4 < \tau_2 \\ CR_{i,g}, & \text{otherwise} \end{cases} \tag{4.9}$$

where rand_j, $j = 1, 2, 3, 4$, are uniform random values on [0, 1], and $\tau_1 = \tau_2 = 0.1$ represent the probabilities to adjust the control parameters.

The newly generated parameter values are used in the mutation and crossover operations to create the corresponding offspring vectors and will replace the previous parameter values if the offspring survive in the selection. It is believed that better parameter values tend to generate individuals which are more likely to survive, and thus the newly generated better values are able to go into the next generation.

Experimental results suggest that jDE performs better than the classic DE/rand/1/bin, DESAP, the adaptive LEP and Best Levy [32], and the FADE algorithm. jDE is further extended by adapting two mutation strategies and the new algorithm, named jDE-2 [22], shows results competitive to the original SaDE (which implements a local Quasi–Newton search procedure after 200 generations) on a set of 25 benchmark functions.

4.2.6 Algorithms Comparison

We have briefly reviewed a number of recently proposed parameter adaptive differential evolution algorithms. Some of them [16], [17] follow a self-adaptive approach (so-called evolution of evolution); i.e., the control parameters themselves go through the mutation, crossover, and selection operations. Some implement a fuzzy logic controller to update parameter values in an adaptive manner [19], [20].

Most adaptive DE algorithms (such as SaDE, jDE, SaNSDE and the JADE algorithm proposed later) follow a *randomization-adaptation* procedure. To be specific, these algorithms randomly generate different control parameter values (mutation factor F and crossover probability CR) according to a certain probability distribution, among which the values leading to successful offspring are recorded and propagated to later generations. According to the results reported in the original papers and a comparison in [22], the randomization-adaptation based DE algorithms usually achieve better performance than other adaptive DE algorithms mentioned above.

The key features of different randomization-adaptation based DE algorithms (including JADE) are summarized in Table 4.1. While adopting the same crossover and

Table 4.1 A summary of the key features of different adaptive differential evolution algorithms

Algorithms	**Features**
SaDE	1. Mutation strategies: DE/rand/1 & DE/current-to-best/2 2. F: Gaussian randomization 3. CR: Gaussian randomization & parameter adaptation
SaNSDE	1. Mutation strategies: DE/rand/1 & DE/current-to-best/2 2. F: Gaussian & Cauchy randomization 3. CR: Gaussian randomization & parameter adaptation
jDE	1. Mutation strategy: DE/rand/1 2. F: Uniform randomization & adaptation 3. CR: Uniform randomization & parameter adaptation
JADE	1. Mutation strategy: Archive-assisted DE/current-to-*p*best/1 or DE/rand-to-*p*best/1 2. F: Cauchy randomization & adaptation 3. CR: Gaussian randomization & parameter adaptation

selection operations, these algorithms differ in their mutation strategies and parameter randomization-adaptation operations. All these adaptive DE algorithms, except JADE, adopt DE/rand/1 as the underlying mutation strategy or one of the two strategies, mainly because it is usually considered less greedy and more robust than other mutation variants. JADE is the only algorithm to implement a relatively greedy mutation strategy which utilizes the information of both the high-quality solutions in the current population and the inferior solutions previously explored during the optimization search. It is expected that a greedy mutation strategy and a parameter adaptation scheme can be mutually beneficial, since the former is capable of increasing the convergence rate while the latter is helpful to maintain the reliability of the algorithm at a high level by adapting to appropriate control parameter values.

The parameter adaptation is composed of two steps: randomization (to generate different parameter values) and adaptation (to select the parameter values that produce successful offspring). jDE generates uniformly distributed parameter values, while all other algorithms randomly generate parameter values according to a (normal or Cauchy) distribution with probability concentrated around the center. SaDE and SaNSDE are the only algorithms that merely randomize the mutation factors without adaptation operations.

Among these algorithms, jDE and JADE introduce the least number of new parameters (τ_1 and τ_2 in jDE, c, p and α in JADE) to update the original problem-dependent control parameter F and CR. The parameters introduced in jDE are shown to be insensitive to optimization problems of different characteristics [22], so are the parameters of JADE as observed in our experimental studies later. SaDE and SaNSDE involve more parameters as described in previous subsections. The sensitivity of these parameters to different problems has not been systematically investigated, whereas the values used in the original papers work well for a variety of benchmark functions.

4.3 JADE: A New Adaptive Differential Evolution Algorithm

In this section, we propose a new parameter adaptive DE algorithm, JADE, that implements a new mutation strategy and updates the control parameters in an adaptive manner. The initialization, crossover and selection operations follow the basic procedure of the classic DE as introduced in Chapter 2.2. We make detailed explanations of the reasons and benefits of the proposed mutation strategy and parameter adaptation operations.

4.3.1 *Initialization*

JADE follows the basic procedure of differential evolution. The initial population $\{\mathbf{x}_{i,0} = (x_{1,i,0}, x_{2,i,0}, \ldots, x_{D,i,0}) | i = 1, 2, \ldots, NP\}$ is randomly generated according to a uniform distribution $x_j^{\mathrm{low}} \leq x_{j,i,0} \leq x_j^{\mathrm{up}}$, for $j = 1, 2, \ldots, D$, where D is the dimension of the problem, NP is the population size, and x_j^{low} and x_j^{low} define the upper and lower limits of the j-th decision variable. After initialization, JADE enters a loop of evolutionary operations (mutation, crossover and selection) and parameter adaptation operations.

4.3.2 *Mutation*

DE/rand/1 is the first mutation strategy developed for DE [1] and is said to be the most successful and widely used scheme in the literature [82]. However, [72] indicates that DE/best/2 may have some advantages over DE/rand/1, and [83] favors DE/best/1 for most technical problems investigated. Also, the authors of [84] argue that the incorporation of the information about the best solution is beneficial and use DE/current-to-best/1 in their algorithm. Compared to DE/rand/*k*, greedy strategies such as DE/current-to-best/*k* and DE/best/*k* benefit from their fast convergence by incorporating best solution information in the evolutionary search. However, the best solution information may also cause problems such as premature convergence due to the resultant decreased population diversity.

In view of the fast but less reliable convergence performance of greedy strategies, a new mutation strategy, named DE/current-to-*p*best, is proposed to serve as the basis of the adaptive DE algorithm JADE proposed in this chapter. As illustrated in Figure 4.1, in DE/current-to-*p*best/1 (without archive), a mutation vector is generated in the following manner:

$$\mathbf{v}_{i,g} = \mathbf{x}_{i,g} + F_i(\mathbf{x}^p_{\mathrm{best},g} - \mathbf{x}_{i,g}) + F_i(\mathbf{x}_{r1,g} - \mathbf{x}_{r2,g}), \tag{4.10}$$

where $\mathbf{x}^p_{\mathrm{best},g}$, named a *p*best solution, is randomly chosen as one of the top $100p\%$ individuals in the current population with $p \in (0, 1]$, $r1$ and $r2$ are distinct integers

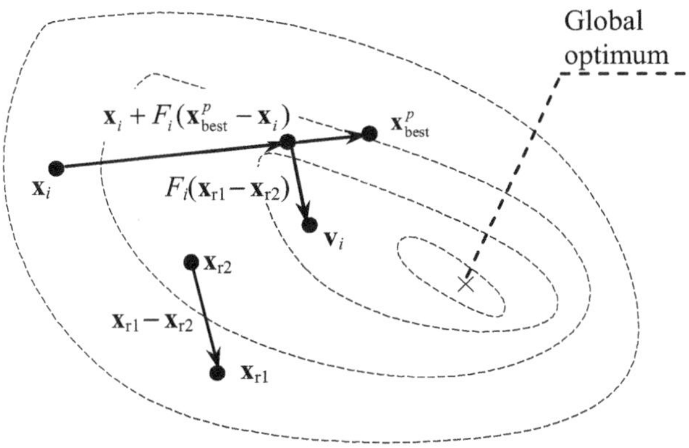

Fig. 4.1 An illustration of the DE/current-to-*p*best/1 mutation strategy (without archive) adopted in JADE. The dashed curves display the contours of the optimization problem. $\mathbf{v}_i$ is the mutation vector generated for individual xi using the associated mutation factor F_i

uniformly chosen from the set $\{1, 2, \ldots, NP\}\backslash\{i\}$, and F_i is the mutation factor that is associated with $\mathbf{x}_i$ and is re-generated at each generation by the adaptation process introduced later in (4.8). DE/current-to-*p*best is indeed a generalization of DE/current-to-best. Any of the top $100p\%$ solutions can be randomly chosen to play the role of the single best solution in DE/current-to-best.

Recently explored inferior solutions, when compared to the current population, provide additional information about the promising progress direction. Denote **A** as the set of archived inferior solutions and **P** as the current population. In archive-assisted DE/current-to-*p*best/1, a mutation vector is generated in the following manner,

$$\mathbf{v}_{i,g} = \mathbf{x}_{i,g} + F_i(\mathbf{x}^p_{\text{best},g} - \mathbf{x}_{i,g}) + F_i(\mathbf{x}_{r1,g} - \tilde{\mathbf{x}}_{r2,g}), \tag{4.11}$$

where $\mathbf{x}_{i,g}$, $\mathbf{x}_{r1,g}$ and $\mathbf{x}^p_{\text{best},g}$are selected from **P** in the same way as in (4.1), and $\tilde{\mathbf{x}}_{r2,g}$ is a vector (distinct from $\mathbf{x}_{i,g}$ and $\mathbf{x}_{r1,g}$)) randomly chosen from the union, $\mathbf{P}\cup\mathbf{A}$, of the archive and current population. The archive operation is made very simple to avoid significant computation overhead. The archive is initiated to be empty. Then, after each generation, we add to the archive the parent solutions that fail in the selection process of (4.5). If the archive size exceeds a certain threshold, say $\alpha \cdot NP$ ($\alpha \geq 0$), then we randomly removed some solutions from the archive to keep the archive size at $\alpha \cdot NP$. It is clear that (4.1) is a special case of (4.2) by setting the archive size to be zero.

The archive provides information about the progress direction and is also capable of improving the diversity of the population. In addition, as shown later in the parameter adaptation operations (4.15) – (4.19), we encourage large random F values that are helpful to increase the population diversity. Thus, although it is biased towards the direction of potential optimum (possibly a local minimum), the proposed archive-assisted DE/current-to-*p*best/1 is not liable to be trapped to a local minimum. As a simple comparison, DE/rand/1 searches a relatively small region that shows no bias to special directions, while DE/current-to-*p*best/1 with archive searches a relatively large region which is biased towards promising progress directions.

To further increase population diversity, a variant of (4.2) (named archive-assisted DE/rand-to-*p*best/1) can be proposed as follows,

$$\mathbf{v}_{i,g} = \mathbf{x}_{r0,g} + F_i(\mathbf{x}^p_{\text{best},g} - \mathbf{x}_{r0,g}) + F_i(\mathbf{x}_{r1,g} - \tilde{\mathbf{x}}_{r2,g}), \tag{4.12}$$

where $\mathbf{x}_{r0,g}$ and $\mathbf{x}_{r1,g}$ ($r0 \neq r1$) are two vectors randomly selected from the population $\mathbf{P}$, $\tilde{\mathbf{x}}_{r2,g}$ is a vector (distinct from $\mathbf{x}_{r0,g}$ and $\mathbf{x}_{r1,g}$) randomly chosen from the union of $\mathbf{P}$ and $\mathbf{A}$. The only difference between (4.11) and (4.12) is that the base vector $\mathbf{x}_{i,g}$ is fixed for each given i in (4.11) while $\mathbf{x}_{r0,g}$ is randomly chosen from the population in (4.12). Denote the size of the archive as $|\mathbf{A}|$. For given i, F_i and $\mathbf{x}^p_{\text{best},g}$, the number of possible mutation vectors is $(NP-1)(|\mathbf{A}|-2)$ according to (4.11), while the number is increased to $NP(NP-1)(|\mathbf{A}|-2)$ according to (4.12). It is helpful to further improve the population diversity during the optimization process.

To avoid confusion, we call the proposed algorithm JADE or JADE+ when the DE/current-to-*p*best in (4.11) or the DE/rand-to-*p*best (4.12) is implemented as the mutation strategy, respectively. In addition, we call the algorithm as "JADE without archive" or "JADE+ without archive" if the archive size is set to be zero.

For simplicity, however, we may simple use JADE to refer to all the variants mentioned above if no confusion is possible.

4.3.3 *Crossover*

After mutation, a 'binomial' crossover operation forms the final trial vector $\mathbf{u}_{i,g} = (u_{1,i,g}, u_{2,i,g}, \ldots, u_{D,i,g})$:

$$u_{j,i,g} = \begin{cases} v_{j,i,g}, & \text{if } \text{rand}_j(0,1) \leq CR_i \quad \text{or} \quad j = j_{\text{rand}} \\ x_{j,i,g}, & \text{otherwise,} \end{cases} \tag{4.13}$$

where rand_j(a,b) is a uniform random number on the interval [a, b] and newly generated for each j, j_{rand} = $\text{randint}_i(1, D)$ is an integer randomly chosen from 1 to D and newly generated for each i, and the crossover probability, $CR_i \in [0, 1]$, roughly corresponds to the average fraction of vector components that are inherited from the mutation vector.

4.3.4 *Selection*

The selection operation selects the better one from the parent vector $\mathbf{x}_{i,g}$ and the trial vector $\mathbf{u}_{i,g}$ according to their fitness values $f(\cdot)$. For example, if we have a minimization problem, the selected vector is given by

$$\mathbf{x}_{i,g+1} = \begin{cases} \mathbf{u}_{i,g}, & \text{if } f(\mathbf{u}_{i,g}) < f(\mathbf{x}_{i,g}) \text{ or } f(\mathbf{u}_{i,g}) = f(\mathbf{x}_{\text{best},g}) \\ \mathbf{x}_{i,g}, & \text{otherwise,} \end{cases} \tag{4.14}$$

and used as a parent vector in the next generation.[1]

The operation in (3.2) is called a *successful update* if the trial vector $\mathbf{u}_{i,g}$ is better than the parent $\mathbf{x}_{i,g}$, i.e., the improvement or evolution progress $\Delta_{i,g} = f(\mathbf{x}_{i,g}) - f(\mathbf{u}_{i,g})$ is positive. Accordingly, the control parameters F_i and CR_i used to generate $\mathbf{u}_{i,g}$ are called the *successful mutation factor* and *successful crossover probability*, respectively.

4.3.5 *Adaptation of* μ_{CR}

In JADE, the adaptation is applied to update the parameters μ_{CR} and μ_F that are used to generate the mutation factor F_i and crossover probability CR_i associated with each individual vector $\mathbf{x}_i$, respectively. The F_i and CR_i are then used to generate the trial vector $\mathbf{u}_i$ when $\mathbf{x}_i$ serves as the base vector in (4.11) and (4.13). The pseudo code of JADE is presented in Table 4.2, where the lines related to DE/current-to-*p*best and the *CR* and *F* adaptations are respectively labeled by "→" and "⇒" for clarity. Detailed adaptive operations are described below.

At each generation g, the crossover probability CR_i of each individual $\mathbf{x}_i$ is independently generated according to a normal distribution $\text{randn}_i(\mu_{CR}, 0.1)$ of mean μ_{CR} and standard deviation 0.1, i.e.,

$$CR_i = \text{randn}_i(\mu_{CR}, 0.1), \tag{4.15}$$

and then truncated to the interval [0, 1].

Denote S_{CR} as the set of all successful crossover probabilities CR_i's at generation g. The mean μ_{CR} is initialized to be 0.5 and then updated at the end of each generation as follows:

$$\mu_{CR} = (1-c)\cdot\mu_{CR} + c\cdot\text{mean}_A(S_{CR}), \tag{4.16}$$

[1] Eq. (4.14) is slightly different from the method usually used in the literature, i.e.,

$$\mathbf{x}_{i,g+1} = \begin{cases} \mathbf{u}_{i,g}, & \text{if } f(\mathbf{u}_{i,g}) < f(\mathbf{x}_{i,g}) \\ \mathbf{x}_{i,g}, & \text{otherwise,} \end{cases}$$

as used in the classic DE, SaDE, and jDE. These two methods have little, if any, difference if the landscape of the optimization function does not contain plane areas. If it contains plane areas (e.g., in the case of the step function f_6 studies in this monograph), the method used in the literature may occasionally cause the optimization process to become stagnant according to our experimental study. The method in Eq. (4.14) is capable of relieving the stagnant problem by allowing an offspring $\mathbf{u}_{i,g}$ to replace its parent if it is better than the parent or if it is equal to the best solution value $f(\mathbf{x}_{\text{best},g})$. In the latter case, at least two solutions ($\mathbf{u}_{i,g}$ and $\mathbf{x}_{\text{best},g}$) achieve the best-so-far function value, indicating that the best-so-far solutions have probably entered a plane area. It is helpful for the algorithm to spread over and pass through the plane area by replacing the parent with its equally good offspring.

Table 4.2 Pseudo code of JADE with archive. The DE/current-to-$p$best operation is denoted by $\rightarrow$, and the adaptive operations by $\Rightarrow$

```
     Procedure of JADE with Archive

     Begin
         Set μ_CR = 0.5; μ_F = 0.5; A = ∅
         Create a random initial population {x_{i,0} | i = 1, 2, ..., NP}
         For g = 1 to G
⇒            S_F = ∅; S_CR = ∅;
             For i = 1 to NP
⇒                Generate CR_i = randn_i(μ_CR, 0.1), F_i = randc_i(μ_F, 0.1)
→                Randomly choose x^p_{best,g} as one of the 100p% best vectors
                 Randomly choose x_{r1,g} ≠ x_{i,g} from current population P
                 Randomly choose x̃_{r2,g} ≠ x_{r1,g} ≠ x_{i,g} from P ∪ A
→                v_{i,g} = x_{i,g} + F_i(x^p_{best,g} − x_{i,g}) + F_i(x_{r1,g} − x̃_{r2,g})
                 Generate j_rand = randint(1, D)
                 For j = 1 to D
                     If j = j_rand or rand(0, 1) ≤ CR_i
                         u_{j,i,g} = v_{j,i,g}
                     Else
                         u_{j,i,g} = x_{j,i,g}
                     End If
                 End For
                 If f(x_{i,g}) < f(u_{i,g}) or f(u_{i,g}) = f(x_{best,g})
⇒                    x_{i,g+1} = u_{i,g}; x_{i,g} → A; CR_i → S_CR, F_i → S_F
                 Else
                     x_{i,g+1} = x_{i,g}
                 End If
             End for
             Randomly remove solutions from A so that |A| ≤ NP
⇒            μ_CR = (1 − c) · μ_CR + c · mean_A(S_CR)
⇒            μ_F = (1 − c) · μ_F + c · mean_L(S_F)
         End for
     End
```

where c is a positive constant between 0 and 1 and $\text{mean}_A(\cdot)$ is the usual arithmetic mean operation.

4.3.6 *Adaptation of* μ_F

At each generation g, the mutation factor F_i of each individual $\mathbf{x}_i$ is independently generated according to a Cauchy distribution $\text{randc}_i(\mu_F, 0.1)$ with location parameter μ_F and scale parameter 0.1,

$$F_i = \text{randc}_i(\mu_F, 0.1), \tag{4.17}$$

and then truncated to be 1 if $F_i \geq 1$ or regenerated if $F_i \leq 0$.

Denote S_F as the set of all successful mutation factors F_i's at generation g. The location parameter μ_F of the Cauchy distribution is initialized to be 0.5 and then updated at the end of each generation as follows:

$$\mu_F = (1-c)\cdot\mu_F + c\cdot\text{mean}_L(S_F), \tag{4.18}$$

where $\text{mean}_L(\cdot)$ is the Lehmer mean:

$$\text{mean}_L(S_F) = \frac{\sum_{F\in S_F} F^2}{\sum_{F\in S_F} F}. \tag{4.19}$$

4.3.7 Explanation of Parameter Adaptation

The adaptation of μ_{CR} is based on the following principle: Better control parameter values tend to generate individuals that are more likely to survive and thus these values should be propagated. The basic operation is thus to record recent successful crossover probabilities and use them to guide the generation of new CR_i's. The standard deviation is set to be relatively small in (4.15) because otherwise the adaptation does not function efficiently; e.g., in the extreme case of an infinite standard derivation, the truncated normal distribution becomes independent of the value of μ_{CR}.

Compared to CR, there are two distinct operations in the adaptation of μ_F. First, F_i's are generated according to a truncated Cauchy distribution. Compared to a normal distribution, the Cauchy distribution is more helpful to diversify the mutation factors and thus avoid premature convergence which frequently occurs in greedy mutation strategies (such as DE/best, DE/current-to-best, and DE/current-to-*p*best) if the mutation factors are highly concentrated around a certain value.

Second, the adaptation of μ_F places more weight on larger successful mutation factors by using the Lehmer mean in (4.19), instead of an arithmetic mean as used in the μ_{CR} adaptation. The Lehmer mean is therefore helpful to propagate larger mutation factors which in turn improve the progress rate. To the contrary, an arithmetic mean of S_F tends to be smaller than the optimal value of the mutation factor and thus leads to a smaller μ_F and causes premature convergence at the end. The decreasing trend is mainly due to the discrepancy between success probability and progress rate of an evolutionary search. Indeed, the DE/current-to-*p*best with small F_i is similar to a $(1+1)$ ES scheme in the sense that both generate an offspring in the small neighborhood of the base vector. For (1+1) ES, it is known that the smaller the mutation variance, usually the higher the successful probability (this is strictly proven for the corridor and sphere functions [11], [12]). However, a mutation variance close to zero obviously leads to a trivial evolution progress. A simple and efficient method is to place more weight on larger successful mutation factors to achieve a faster progress rate.

The randomization process of F in (4.17) is also helpful to improve the population diversity. According to the analytical results in the Appendix of Section 4.8, the variance of offspring vectors $\mathbf{u}$ can be written as a function of the variance of parent vectors $\mathbf{x}$,

$$Var(\mathbf{u}) = < 1 - 2E(CR) + E(2 - 2F + 4F^2)\Phi, Var(x) >, \qquad (4.20)$$

as far as only the mutation and crossover of JADE are concerned. In (4.20), <A, B> denotes the element-wise product of two matrices A and B, and $\Phi = \{\phi_{j,k}\}$ is a matrix of positive constants $\phi_{k,k} = E(CR)$ and $\phi_{j,k} = E(CR^2)$, for $j \neq k$. Consider that $E(F^2) = E(F)^2 + Var(F)$ for a random variable F. It is clear that, compared to a constant F (i.e., $Var(F) = 0$), the randomization process tends to improve the variance $Var(\mathbf{u})$ of the offspring vectors and thus increase the population diversity, which in turn is helpful to avoid false convergence to local minima.

Regarding the constant c in (4.16) and (4.18), no adaptation takes place if $c = 0$. Otherwise, the *life span* of a successful CR_i or F_i is roughly $1/c$ generations: after $1/c$ generations, the old value of μ_{CR} or μ_F is reduced by a factor $1/e \approx 0.37$ in a new update.

4.3.8 Discussion of Parameter Settings

The classic DE has two control parameters F and CR that need to be adjusted by the user. These parameters are known to be problem-dependent and thus tedious trial and error is usually required to find their appropriate values for each specific problem. JADE, on the other hand, takes advantages of the problem insensitive property of its parameters c and p whose values can be appropriately chosen based on their role in the algorithm. As shown by simulations, JADE usually performs best with $1/c \in [5, 20]$ and $p \in [5\%, 20\%]$; i.e., the life span of μ_{CR} and μ_F values ranges from five to twenty generations, and we consider the top 5% – 20% high-quality solutions in the mutation operation.

4.3.9 Algorithm Complexity

In general differential evolution algorithms including JADE, there are $O(NP \cdot D)$ arithmetic operations in both mutation and crossover. The complexity of finding all $p$best solutions is $O(NP \cdot \log(NP))$ in JADE. In addition, both selection and parameter adaptation take $O(NP)$ operations. The total complexity of JADE is therefore $O(G \cdot NP \cdot (D + \log(NP)))$, where G is the total number of generations. In the literature of differential evolution, it is usually recommended to set the population size NP to be proportional to the problem dimension D. Thus, the total complexity of JADE is $O(G \cdot D^2)$, which is the same as the classic DE and many other DE algorithms.

4.4 Performance Analysis for Low- to Moderate-Dimensional Problems

In this section, JADE is applied to minimize a set of 13 benchmark functions of dimensions $D = 30$ or 100 and a set of Dixon–Szegö functions of lower dimension

$D = 2 \sim 6$, as shown in Table 4.3 and Table 4.4. The scalability of JADE as well as JADE+ for higher dimensional problems will be analyzed later in Chapter 4.6.

JADE is compared with two recent adaptive DE algorithms jDE and SaDE, the classic DE/rand/1/bin, and a canonical PSO algorithm [31] . It is also compared with rand-JADE and nona-JADE, two variants of JADE, to identify the contributions from different components of JADE. For fair comparison, we set the parameters of JADE to be fixed, $p = 0.05$ and $c = 0.1$, in all simulations. We follow the parameter settings in the original paper of jDE [22] and SaDE [21], except that the Quasi-Newton local search is disabled in SaDE. The parameters of PSO are chosen from [31] as they work better than the values proposed in other literature we studied. The parameters of DE/rand/1/bin are set to be $F = 0.5$ and $CR = 0.9$, as used or recommended in [1], [9], [10], [22].

Summarized in Table 4.3 are the 13 scalable benchmark functions that are widely used in the literature [85]. While all these functions have an optimal value $f^* = 0$, some different characteristics are briefly summarized as follows. $f_1 - f_4$ are continuous unimodal functions. f_5 is the Rosenbrock function which is unimodal for $D = 2$

Table 4.3 Test functions of dimension D. Each of them has a global minimum value of 0. These are sphere, Schwefel 2.22, Schwefel 1.2, Schwefel 2.21, Rosenbrock, step, noisy quartic, Schwefel 2.26, Rastrigin, Ackley, Griewank, and two penalized functions, respectively [85]

Test Functions	Initial Range
$f_1(x) = \sum_{i=1}^{D} x_i^2$	$[-100, 100]^D$
$f_2(x) = \sum_{i=1}^{D} \lvert x_i\rvert + \prod_{i=1}^{D} \lvert x_i\rvert$	$[-10, 10]^D$
$f_3(x) = \sum_{i=1}^{D} (\sum_{j=1}^{i} x_j)^2$	$[-100, 100]^D$
$f_4(x) = \max_i\{\lvert x_i\rvert\}$	$[-100, 100]^D$
$f_5(x) = \sum_{i=1}^{D-1} \left[100(x_{i+1} - x_i^2)^2 + (x_i - 1)^2\right]$	$[-30, 30]^D$
$f_6(x) = \sum_{i=1}^{D} \lfloor x_i + 0.5 \rfloor^2$	$[-100, 100]^D$
$f_7(x) = \sum_{i=1}^{D} i x_i^4 + rand[0, 1)$	$[-1.28, 1.28]^D$
$f_8(x) = \sum_{i=1}^{D} -x_i \sin\sqrt{\lvert x_i\rvert} + D \cdot 418.98288727243369$	$[-500, 500]^D$
$f_9(x) = \sum_{i=1}^{D} [x_i^2 - 10\cos(2\pi x_i) + 10]$	$[-5.12, 5.12]^D$
$f_{10}(x) = -20\exp(-0.2\sqrt{\frac{1}{D}\sum_{i=1}^{D} x_i^2}) - \exp\left(\frac{1}{D}\sum_{i=1}^{D}\cos(2\pi x_i)\right) + 20 + e$	$[-32, 32]^D$
$f_{11}(x) = \frac{1}{4000}\sum_{i=1}^{D} x_i^2 - \prod_{i=1}^{D}\cos(x_i/\sqrt{i}) + 1$	$[-600, 600]^D$
$f_{12}(x) = \frac{\pi}{D}\{10\sin^2(\pi y_1) + \sum_{i=1}^{D-1}(y_i - 1)^2[1 + 10\sin^2(\pi y_{i+1})] + (y_D - 1)^2\} + \sum_{i=1}^{D} u(x_i, 10, 100, 4)$	$[-50, 50]^D$
$f_{13}(x) = 0.1\{\sin^2(3\pi x_1) + \sum_{i=1}^{D-1}(x_i - 1)^2[1 + \sin^2(3\pi x_{i+1})] + (x_D - 1)^2[1 + \sin^2(2\pi x_D)]\} + \sum_{i=1}^{D} u(x_i, 5, 100, 4)$	$[-50, 50]^D$

In f_{12} and f_{13}, $y_i = 1 + \frac{1}{4}(x_i + 1)$ and $u(x_i, a, k, m) = \begin{cases} k(x_i - a)^m, & x_i > a \\ 0, & -a \le x_i \le a \\ k(-x_i - a)^m, & x_i < -a \end{cases}$

Table 4.4 Test functions of dimension D. Each of them has a global minimum value of 0. These are sphere, Schwefel 2.22, Schwefel 1.2, Schwefel 2.21, Rosenbrock, step, noisy quartic, Schwefel 2.26, Rastrigin, Ackley, Griewank, and two penalized functions, respectively [87]

Functions	D	Initial Range	Number of Local Min	Number of Global Min	Approx. Global Min
f_{14}: Branin	2	$[-5,10]\times[0, 15]$	3	3	0.397887
f_{15}: Goldstein-Price	2	$[-2, 2]^2$	4	1	3.00000
f_{16}: Hartman3	3	$[0, 1]^3$	4	1	-3.86278
f_{17}: Hartman6	6	$[0, 1]^6$	4	1	-3.32237
f_{18}: Shekel5	4	$[0, 10]^4$	5	1	-10.1532
f_{19}: Shekel7	4	$[0, 10]^4$	7	1	-10.4029
f_{20}: Shekel10	4	$[0, 10]^4$	10	1	-10.5364

and 3 but has multiple minima in higher dimensional cases [83]. f_6 is a discontinuous step function, and f_7 is a noisy quartic function. $f_8 - f_{13}$ are multimodal and the number of their local minima increases exponentially with the problem dimension [32] [33]. In addition, f_8 is the only bound-constrained function investigated in this chapter. The characteristics of the multimodal Dixon–Szegö functions are briefly described in Table 4.4. Interested readers are referred to [87] for details.

In all simulations, we set the population size NP to be 30, 100, and 400 in the cases of $D \leq 10$, $= 30$ and $= 100$, respectively. In addition, all results reported in this section are calculated based on 50 independent runs. For clarity, the results of the best and second best algorithms are generally marked in **boldface** and *italic*, respectively.

4.4.1 Comparison of JADE with Other Evolutionary Algorithms

For $f_1 - f_{13}$, the mean and standard deviation of the best-so-far function values are summarized in Table 4.5 and Table 4.6. These statistics are calculated for $f_1 - f_4$, and f_7 only at the end of the optimization. For other functions, intermediate results are also reported because the final results obtained by different algorithms might be identical to zero or have a certain residual error (see the error floor of f_{10} and f_{13} in Figure 4.3) due to precision problems of MATLAB. [88] In these cases, only the intermediate results are compared and may be marked in boldface or italic.

In Table 4.7 and Table 4.8, we summarize the *s*uccess *r*ate (SR) of each algorithm and the average number of *f*unction *e*valuation*s* over *s*uccessful runs (FESS). An experiment is considered as successful if the best solution is found with sufficient accuracy: 10^{-2} for the noisy function f_7 and 10^{-8} for all others. FESS and SR are useful for comparing the convergence rate (in successful runs) and the reliability of algorithms, respectively.

For convenience of illustration, we plot the convergence graph for some 30- and 100-dimensional problems in Figure 4.2 and Figure 4.3, respectively. Note that in

Table 4.5 The mean and standard deviation (in parenthesis) of the experimental results of 30-dimensional problems $f_1 - f_{13}$. The results are obtained based on 50 independent runs

	Gen	JADE w/o archive	JADE with archive	jDE	SaDE	DE/rand /1/bin	PSO
f_1	1500	**1.8E-60**	*1.3E-54*	2.5E-28	4.5E-20	9.8E-14	9.6E-42
		(8.4E-60)	*(9.2E-54)*	(3.5E-28)	(6.9E-20)	(8.4E-14)	(2.7E-41)
f_2	2000	**1.8E-25**	3.9E-22	*1.5E-23*	1.9E-14	1.6E-09	9.3E-21
		(8.8E-25)	(2.7E-21)	*(1.0E-23)*	(1.1E-14)	(1.1E-09)	(6.3E-20)
f_3	5000	*5.7E-61*	**6.0E-87**	5.2E-14	9.0E-37	6.6E-11	2.5E-19
		(2.7E-60)	**(1.9E-86)**	(1.1E-13)	(5.4E-36)	(8.8E-11)	(3.9E-19)
f_4	5000	*8.2E-24*	**4.3E-66**	1.4E-15	7.4E-11	4.2E-01	4.4E-14
		(4.0E-23)	**(1.2E-65)**	(1.0E-15)	(1.82E-10)	(1.1E+00)	(9.3E-14)
f_5	3000	**8.0E-02**	*3.2E-01*	1.3E+01	2.1E+01	2.1E+00	2.5E+01
		(5.6E-01)	*(1.1E+00)*	(1.4E+01)	(7.8E+00)	(1.5E+00)	(3.2E+01)
	20000	8.0E-02	3.2E-01	8.0E-02	1.8E+01	8.0E-02	1.7E+01
		(5.6E-01)	(1.1E+00)	(5.6E-01)	(6.7E+00)	(5.6E-01)	(2.3E+01)
f_6	100	**2.9E+00**	*5.6E+00*	1.0E+03	9.3E+02	4.7E+03	4.5E+01
		(1.2E+00)	*(1.6E+00)*	(2.2E+02)	(1.8E+02)	(1.1E+03)	(2.4E+01)
	1500	0.0E+00	0.0E+00	0.0E+00	0.0E+00	0.0E+00	8.0E-02
		(0.0E+00)	(0.0E+00)	(0.0E+00)	(0.0E+00)	(0.0E+00)	(2.7E-01)
f_7	3000	**6.4E-04**	*6.8E-04*	3.3E-03	4.8E-03	4.7E-03	2.5E-03
		(2.5E-04)	*(2.5E-04)*	(8.5E-04)	(1.2E-03)	(1.2E-03)	(1.4E-03)
f_8	1000	*3.3E-05*	7.1E+00	**7.9E-11**	4.7E+00	5.9E+03	2.4E+03
		(2.3E-05)	(2.8E+01)	**(1.3E-10)**	(3.3E+01)	(1.1E+03)	(6.7E+02)
	9000	0.0E+00	7.1E+00	0.0E+00	4.7E+00	5.7E+01	2.4E+03
		(0.0E+00)	(2.8E+01)	(0.0E+00)	(3.3E+01)	(7.6E+01)	(6.7E+02)
f_9	1000	**1.0E-04**	*1.4E-04*	1.5E-04	1.2E-03	1.8E+02	5.2E+01
		(6.0E-05)	*(6.5E-05)*	(2.0E-04)	(6.5E-04)	(1.3E+01)	(1.6E+01)
	5000	0.0E+00	0.0E+00	0.0E+00	0.0E+00	7.1E+01	5.2E+01
		(0.0E+00)	(0.0E+00)	(0.0E+00)	(0.0E+00)	(2.1E+01)	(1.6E+01)
f_{10}	500	**8.2E-10**	*3.0E-09*	3.5E-04	2.7E-03	1.1E-01	4.6E-01
		(6.9E-10)	*(2.2E-09)*	(1.0E-04)	(5.1E-04)	(3.9E-02)	(6.6E-01)
	2000	4.4E-15	4.4E-15	4.7E-15	4.3E-14	9.7E-11	4.6E-01
		(0.0E+00)	(0.0E+00)	(9.6E-16)	(2.6E-14)	(5.0E-11)	(6.6E-01)
f_{11}	500	**9.9E-08**	2.0E-04	*1.9E-05*	7.8E-04)	2.0E-01	1.3E-02
		(6.0E-07)	(1.4E-03)	*(5.8E-05)*	(1.2E-03	(1.1E-01)	(1.7E-02)
	3000	0.0E+00	2.0E-04	0.0E+00	0.0E+00	0.0E+00	1.1E-02
		(0.0E+00)	(1.4E-03)	(0.0E+00)	(0.0E+00)	(0.0E+00)	(1.6E-02)
f_{12}	500	**4.6E-17**	*3.8E-16*	1.6E-07	1.9E-05	1.2E-02	1.9E-01
		(1.9E-16)	*(8.3E-16)*	(1.5E-07)	(9.2E-06)	(1.0E-02)	(3.9E-01)
	1500	1.6E-32	1.6E-32	2.6E-29	1.2E-19	1.1E-14	1.9E-01
		(5.5E-48)	(5.5E-48)	(7.5E-29)	(2.0E-19)	(1.0E-14)	(3.9E-01)
f_{13}	500	**2.0E-16**	*1.2E-15*	1.5E-06	6.1E-05	7.5E-02	2.9E-03
		(6.5E-16)	*(2.8E-15)*	(9.8E-07)	(2.0E-05)	(3.8E-02)	(4.8E-03)
	1500	1.4E-32	1.4E-32	1.9E-28	1.7E-19	7.5E-14	2.9E-03
		(1.1E-47)	(1.1E-47)	(2.2E-28)	(2.4E-19)	(4.8E-14)	(4.8E-03)

Table 4.6 The mean and standard deviation (in parenthesis) of the experimental results of 100-dimensional problems $f_1 - f_{13}$. The results are obtained based on 50 independent runs

	Gen	JADE w/o archive	JADE with archive	jDE	SaDE	DE/rand /1/bin	PSO
f_1	2000	*1.2E-48*	**5.4E-67**	5.0E-15	2.9E-08	3.5E+01	6.0E-11
		(1.5E-48)	**(1.6E-66)**	(1.7E-15)	(3.2E-08)	(9.6E+00)	(2.5E-10)
f_2	3000	*1.1E-41*	**9.2E-51**	4.1E-15	1.7E-05	2.3E+00	2.8E-04
		(5.1E-41)	**(2.2E-50)**	(1.1E-15)	(3.8E-06)	(5.6E-01)	(1.3E-03)
f_3	8000	*1.2E-26*	**2.2E-37**	5.4E-02	2.4E-13	2.1E+05	1.2E+02
		(2.0E-26)	**(2.5E-37)**	(2.7E-02)	(5.2E-13)	(3.1E+04)	(6.7E+01)
f_4	15000	1.9E-02	**3.2E-71**	*3.1E-09*	1.1E+00	9.3E+01	4.9E+01
		(1.5E-02)	**(8.3E-71)**	*(5.9E-10)*	(4.0E-01)	(2.8E+00)	(2.5E+01)
f_5	6000	*5.6E-01*	**4.0E-01**	7.2E+01	9.4E+01	9.5E+01	1.3E+02
		(1.4E+00)	**(1.2E+00)**	(1.1E+01)	(4.0E-01)	(1.4E+01)	(4.8E+01)
	20000	5.6E-01	4.0E-01	9.1E-03	9.1E+01	4.3E+01	6.3E+01
		(1.4E+00)	(1.2E+00)	(2.5E-02)	(3.1E-01)	(1.2E+01)	(4.2E+01)
f_6	100	**1.1E+02**	*1.2E+02*	7.1E+04	3.3E+04	1.8E+05	1.9E+04
		(1.5E+01)	*(1.3E+01)*	(6.0E+03)	(2.1E+03)	(1.8E+04)	(4.5E+03)
	1500	1.6E-01	0.0E+00	0.0E+00	0.0E+00	3.2E+02	4.7E+01
		(3.7E-01)	(0.0E+00)	(0.0E+00)	(0.0E+00)	(6.3E+01)	(7.9E+01)
f_7	6000	*1.1E-03*	**7.8E-04**	8.1E-03	1.0E-02	2.9E-02	9.2E-03
		(2.1E-04)	**(1.4E-04)**	(9.0E-04)	(4.9E-03)	(5.7E-03)	(2.6E-03)
f_8	1000	8.9E+03	8.6E+03	**4.9E+03**	*5.4E+03*	3.2E+04	9.5E+03
		(3.0E+02)	(4.2E+02)	**(4.1E+02)**	*(3.7E+02)*	(4.7E+02)	(1.3E+03)
	9000	1.1E-10	1.1E-10	1.1E-10	1.1E-10	3.0E+04	9.4E+03
		(0.0E+00)	(0.0E+00)	(0.0E+00)	(0.0E+00)	(9.0E+02)	(1.2E+03)
f_9	3000	1.9E-01	2.0E-01	**2.1E-04**	*9.1E-03*	8.6E+02	3.4E+02
		(3.8E-02)	(3.7E-02)	**(2.1E-04)**	*(1.8E-03)*	(2.2E+01)	(4.4E+01)
	9000	0.0E+00	0.0E+00	0.0E+00	0.0E+00	8.1E+02	3.4E+02
		(0.0E+00)	(0.0E+00)	(0.0E+00)	(0.0E+00)	(1.8E+01)	(4.4E+01)
f_{10}	500	*7.9E-06*	**4.2E-07**	8.5E-01	1.6E+00	1.5E+01	3.6E+00
		(2.6E-06)	**(1.2E-07)**	(1.2E-01)	(1.2E-01)	(5.8E-01)	(9.3E-01)
	3000	8.9E-15	8.0E-15	9.9E-14	2.1E-07	1.3E-01	2.6E+00
		(2.1E-15)	(0.0E+00)	(2.0E-14)	(1.0E-07)	(2.4E-02)	(6.8E-01)
f_{11}	500	*3.9E-04*	**1.5E-04**	1.1E+00	1.1E+00	2.7E+02	1.0E+00
		(2.0E-03)	**(1.0E-03)**	(2.0E-02)	(1.8E-02)	(4.4E+01)	(5.6E-01)
	3000	3.9E-04	1.5E-04	0.0E+00	8.6E-13	2.0E-01	8.8E-02
		(2.0E-03)	(1.0E-03)	(0.0E+00)	(8.2E-13)	(5.8E-02)	(2.5E-01)
f_{12}	500	*2.2E-11*	**2.8E-13**	4.0E+00	2.4E+00	1.8E+09	1.1E+01
		(1.2E-11)	**(9.8E-13)**	(6.8E-01)	(3.9E-01)	(5.1E+08)	(3.4E+00)
	3000	4.7E-33	4.7E-33	1.7E-25	1.4E-11	1.7E+00	1.3E-01
		(2.2E-34)	(6.8E-49)	(7.7E-26)	(9.1E-12)	(1.5E+00)	(1.6E-01)
f_{13}	500	*1.9E-09*	**5.8E-12**	3.1E+01	1.2E+01	2.4E+09	9.8E+01
		(1.5E-09)	**(5.5E-12)**	(7.8E+00)	(1.8E+00)	(1.1E+09)	(2.4E+01)
	3000	1.4E-32	1.4E-32	2.1E-24	1.3e-11	5.1E+00	1.4E-01
		(1.1E-47)	(1.1E-47)	(1.5E-24)	(9.6E-12)	(3.2E+00)	(5.6E-01)

Table 4.7 Experimental results of 30-dimensional problems $f_1 - f_{13}$, averaged over 50 independent runs. Note that the best and the second best among JADE and other competitive algorithms (except JADE's two variants) are marked in boldface and italic, respectively

Functions		f_1	f_2	f_3	f_4	f_5	f_6	f_7
JADE w/o	SR	100	100	100	100	98	100	100
archive	FESS	**2.9E+4**	**5.2E+4**	*9.4E+4*	*1.7E+5*	*1.5E+5*	**1.1E+4**	**2.9E+4**
JADE with	SR	100	100	100	100	96	100	100
archive	FESS	*3.0E+4*	*5.6E+4*	**7.7E+4**	**7.4E+4**	**1.1E+5**	*1.2E+4*	*3.1E+4*
jDE	SR	100	100	100	100	98	100	100
	FESS	6.0E+4	8.3E+4	3.4E+5	3.0E+5	5.8E+5	2.3E+4	1.0E+5
SaDE	SR	100	100	100	100	24	100	100
	FESS	7.3E+4	1.2E+5	1.8E+5	2.9E+5	2.8E+5	2.7E+4	1.3E+5
DE/rand	SR	100	100	100	6	98	100	100
/1/bin	FESS	1.1E+5	1.9E+5	4.2E+5	3.5E+5	4.4E+5	4.2E+4	1.5E+5
PSO	SR	100	100	100	100	4	92	100
	FESS	4.0E+4	6.7E+4	2.6E+5	3.1E+5	1.2E+6	2.4E+4	7.4E+4
rand-JADE	SR	100	100	38	66	0	100	84
(w/o archive)	FESS	1.2E+5	1.9E+5	2.9E+5	3.2E+5	–	3.6E+4	2.1E+5
nona-JADE	SR	100	100	100	0	88	100	100
(w/o archive)	FESS	2.8E+4	4.7E+4	2.4E+5	–	4.7E+5	1.1E+4	3.1E+4

Functions		f_8	f_9	f_{10}	f_{11}	f_{12}	f_{13}
JADE	SR	100	100	100	100	100	100
(w/o archive)	FESS	1.3E+5	*1.3E+5*	**4.5E+4**	**3.3E+4**	**2.7E+4**	**3.0E+4**
JADE	SR	94	100	100	100	100	100
(with archive)	FESS	1.3E+5	*1.3E+5*	*4.7E+4*	*3.7E+4*	*2.9E+4*	*3.1E+4*
jDE	SR	100	100	100	100	100	100
	FESS	**8.9E+4**	**1.2E+5**	9.1E+4	6.3E+4	5.5E+4	6.0E+4
SaDE	SR	100	100	100	100	100	100
	FESS	*1.2E+5*	1.7E+5	1.2E+5	8.0E+4	7.1E+4	7.4E+4
DE/rand n	SR	60	0	100	100	100	100
/1/bi	FESS	3.5E+5	–	1.7E+5	1.1E+5	9.9E+4	1.1E+5
PSO	SR			66	36	62	74
	FESS	–	–	6.5E+4	4.3E+4	4.8E+4	4.3E+4
rand-JADE	SR	100	100	84	100	100	100
(w/o archive)	FESS	1.6E+5	1.6E+5	2.0E+5	1.4E+5	1.1E+5	1.2E+5
nona-JADE	SR	16	0	100	100	100	100
(w/o archive)	FESS	2.3E+5	–	4.4E+4	3.0E+4	2.5E+4	2.8E+4

these graphs we plot the curves of median values (instead of the mean values reported in the tables), because these curves, together with box-and-whisker diagrams, provide more information when an algorithm may lead to false convergence only occasionally. The box-and-whisker diagrams are plotted at certain generations for the best and second best algorithms. It is helpful to illustrate the spread of results over 50 independent runs.

Table 4.8 Experimental results of 100-dimensional problems $f_1 - f_{13}$, averaged over 50 independent runs. Note that the best and the second best among JADE and other competitive algorithms (except JADE's two variants) are marked in boldface and italic, respectively

Functions		f_1	f_2	f_3	f_4	f_5	f_6	f_7
JADE	SR	100	100	100	0	86	84	100
(w/o archive)	FESS	*2.0E+5*	*2.9E+5*	*1.3E+6*	–	*2.2E+6*	*8.8E+4*	*2.3E+5*
JADE	SR	100	100	100	100	90	100	100
(with archive)	FESS	**1.6E+5**	**2.7E+5**	**9.6E+5**	**7.7E+5**	**1.5E+6**	**6.2E+4**	**2.0E+5**
jDE	SR	100	100	0	100	2	100	96
	FESS	5.5E+5	7.6E+5	–	*5.7E+6*	7.7E+6	2.2E+5	2.0E+6
SaDE	SR	36	0	100	0	0	100	54
	FESS	7.8E+5	–	2.1E+6	–	–	2.6E+5	1.6E+6
DE/rand	SR							
/1/bin	FESS	–	–	–	–	–	–	–
PSO	SR	100	10	0	0	0	2	70
	FESS	6.2E+5	1.1E+6	–	–	–	5.0E+5	1.9E+6
rand-JADE	SR						74	0
(w/o archive)	FESS	–	–	–	–	–	5.8E+5	–
nona-JADE	SR	100	100	0	0	90	100	100
(w/o archive)	FESS	2.0E+5	3.3E+5	–	–	5.0E+6	7.3E+4	3.5E+5

Functions		f_8	f_9	f_{10}	f_{11}	f_{12}	f_{13}
JADE	SR	100	100	100	96	100	100
(w/o archive)	FESS	*1.3E+6*	*1.5E+6*	*2.9E+5*	*1.9E+5*	*1.6E+5*	*1.9E+5*
JADE	SR	100	100	100	98	100	100
(with archive)	FESS	1.4E+6	1.5E+6	**2.4E+5**	**1.7E+5**	**1.4E+5**	**1.6E+5**
jDE	SR	100	100	100	100	100	100
	FESS	**9.5E+5**	**1.4E+6**	8.0E+5	5.4E+5	5.7E+5	5.8E+5
SaDE	SR	100	100	0	100	100	100
	FESS	1.6E+6	1.8E+6	–	8.0E+5	9.2E+5	9.2E+5
DE/rand	SR						
/1/bin	FESS	–	–	–	–	–	–
PSO	SR			2	36	28	54
	FESS	–	–	1.1E+6	6.0E+5	9.5E+5	8.7E+5
rand-JADE	SR	100	100	0	0	0	0
(w/o archive)	FESS	2.5E+6	2.5E+6	–	–	–	–
nona-JADE	SR			100	100	100	100
(w/o archive)	FESS	–	–	3.0E+5	2.0E+5	1.7E+5	1.9E+5

Important observations about the rate and reliability of different algorithms can be made from the results presented in Table 4.5 – Table 4.8, and Figure 4.3 – Figure 4.4. First, these results suggest that the overall convergence rate of JADE with archive and without archive are the best and second best algorithms for the set of problems $f_1 - f_{13}$: JADE without archive works best in the relatively low dimensional case ($D = 30$), while JADE with archive achieves the best performance

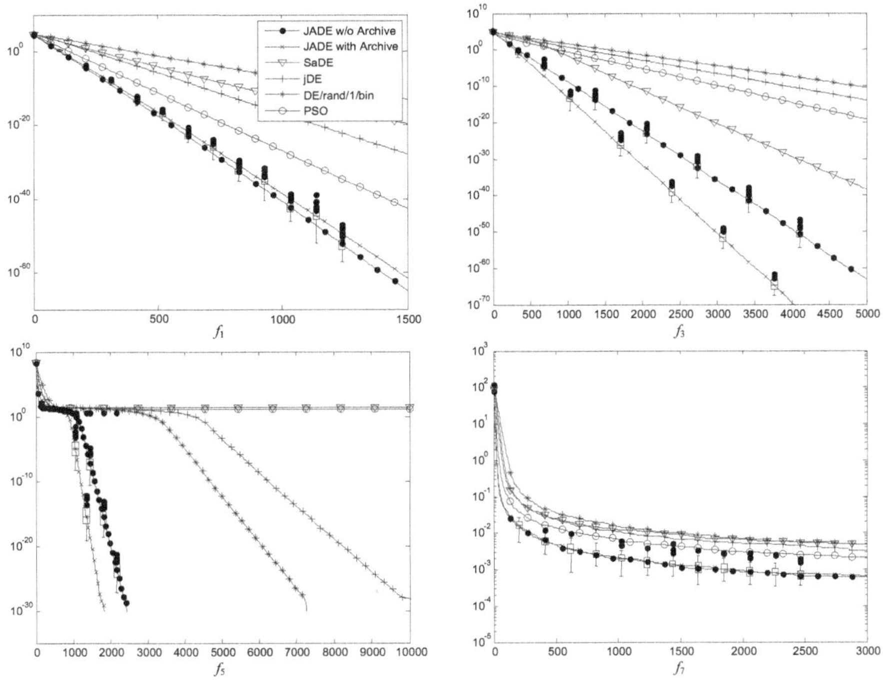

Fig. 4.2 Convergence graph for test functions f_1, f_3, f_5, and f_7 ($D = 30$). The horizontal axis is the number of generations, and the vertical axis is the median of function values over 50 independent runs. The box-and-whisker diagrams are plotted for both JADE and the best one among other algorithms

in the high dimensional case (D = 100). This may be because a population size of 400 is not sufficient for various 100 dimensional problems, while the archive can relieve this problem to some extent by introducing more diversity in the mutation operation. In both cases (with and without archive), JADE converges fastest for the optimization of $f_1 - f_7$ and $f_{10} - f_{13}$, which are either unimodal or multimodal, clean or noisy, continuous or discontinuous. In the cases of f_8 and f_9, jDE performs best and JADE achieves very competitive results. The convergence rates of SaDE and PSO usually rank the fourth or fifth and are better than DE/rand/1/bin.

Second, it is important to compare the reliability of different algorithms, as shown in Table 4.7 and Table 4.8. PSO and DE/rand/1/bin performs worst because the former suffers from premature convergence while the latter usually converges very slowly within a limited number of function evaluations. SaDE is usually better than PSO and DE/rand/1/bin but is clearly less satisfactory for some problems especially when the dimension is high. jDE, as expected, is shown to work very well due to its underlying mutation strategy 'DE/rand/1' which is more robust than greedy strategies. It is interesting that JADE shows high reliability similar to jDE for all 13 high-dimensional functions. The high reliability and fast convergence rate stem both from the adaptive parameter control, and from the direction information and

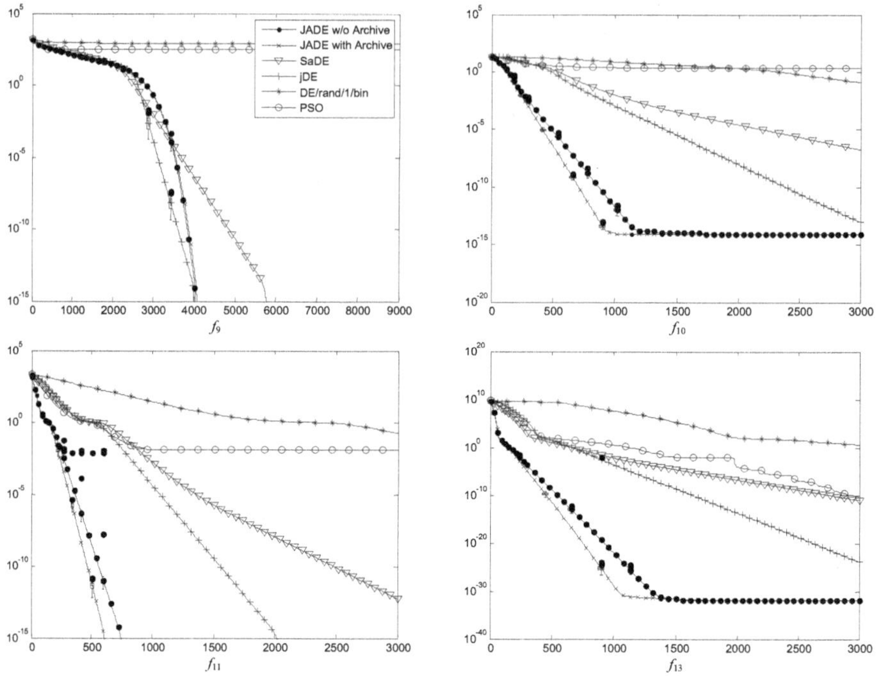

Fig. 4.3 Convergence graph for test functions f_9, f_{10}, f_{11}, and f_{13} (D = 100). The horizontal axis is the number of generations, and the vertical axis is the median of function values over 50 independent runs. The box-and-whisker diagrams are plotted for both JADE and the best one among other algorithms

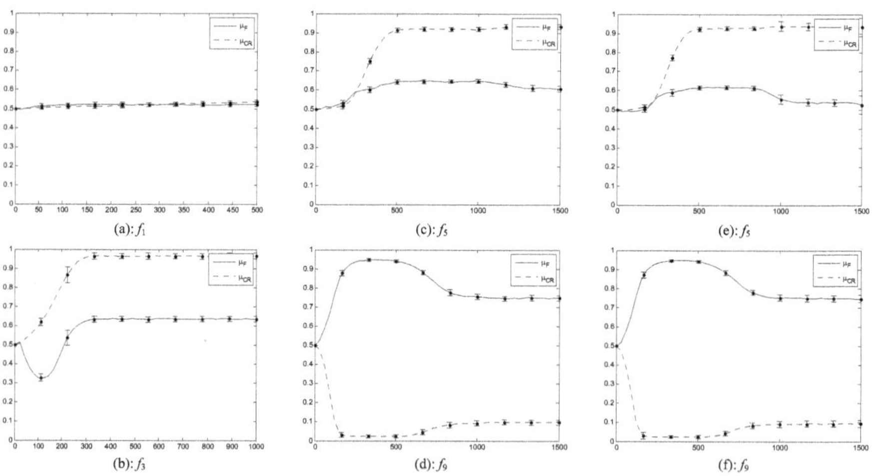

Fig. 4.4 The evolutions of μ_F and μ_{CR} in JADE in the optimization of f_1, f_3, f_5, and f_9 (D = 30). (a) – (d): JADE without archive; (e) – (f): JADE with archive. The horizontal axis is the number of generations. The curve shows the mean and the standard error of μ_F and μ_{CR} calculated based on 50 independent runs

diversity improvement mechanism introduced by the 'current-to-*p*best/1' mutation strategy and the archive operation.

In the case of low-dimensional Dixon–Szegö functions, simulation results in Table 4.9 show that there is no obviously superior algorithm in terms of the convergence performance. This is similar to the observations in the literature [22], where adaptive DE algorithms do not show obvious performance improvement over their non-adaptive counterparts for this set of functions. These observations are not surprising after further simulation investigation. First, various simulations of DE/rand/1/bin with F and CR chosen from $\{0.1, 0.3, 0.5, 0.7, 0.9\}$ show that the parameter values "$F = 0.5$, $CR = 0.9$" consistently achieve the best or near best performance for all Dixon–Szegö functions. Second, simulation results also indicate that parameter adaptation does not function efficiently within the small number of generations required for the optimization of these low dimensional problems.

Table 4.9 The mean and standard deviation (in parenthesis) of the experimental results of the Dixon-Szegö functions $f_{14} - f_{20}$. The results are obtained based on 50 independent runs

	Gen	JADE w/o archive	JADE with archive	jDE	SaDE	DE/rand/1/bin	PSO
f_{14}	200	0.397887	0.397887	0.397887	0.397887	0.397887	0.397887
		(0.0E+00)	(0.0E+00)	(0.00E+00)	(1.4E-13)	(0.0E+00)	(0.0E+00)
f_{15}	200	3.00000	3.00000	3.00000	3.00000	3.00000	3.00000
		(1.1E-15)	(5.4E-16)	(3.44E-16)	(3.0E-15)	(6.4E-16)	(1.3E-15)
f_{16}	200	-3.86278	-3.86278	-3.86278	-3.86278	-3.86278	-3.86278
		(0.0E+00)	(1.3E-16)	(0.00E+00)	(3.1E-15)	(0.0E+00)	(6.2E-17)
f_{17}	200	-3.31044	-3.30806	-3.27707	-3.31425	-3.24608	-3.25800
		(3.6E-02)	(3.9E-02)	(5.79E-02)	(2.8E-02)	(5.7E-02)	(5.9E-02)
f_{18}	200	-10.1532	-9.54691	-10.1532	-10.0522	-10.1532	-5.79196
		(4.0E-14)	(1.6E+00)	(1.60E-12)	(7.1E-01)	(4.2E-16)	(3.3E+00)
f_{19}	200	-10.4029	-10.4029	-10.4029	-10.4029	-10.4029	-7.14128
		(9.4E-16)	(2.4E-12)	(2.77E-15)	(6.6E-11)	(2.5E-16)	(3.5E+00)
f_{20}	200	-10.5364	-10.5364	-10.5364	-10.5363	-10.5364	-7.02213
		(8.1E-12)	(6.1E-14)	(6.11E-16)	(9.2E-04)	(7.1E-16)	(3.6E+00)

In Table 4.10, JADE is further compared with the reported results of other evolutionary algorithms in the literature: the adaptive LEP and Best Levy algorithms in [33, Table III], NSDE in [89, Tables 1 – 3] and jDE in [22, Table III]. It is clear that JADE achieves overall better performance than other competitive algorithms for the 9 test functions investigated. Similar to our previous observations, JADE is better than jDE in all the cases except f_8 and f_9 where the results of JADE are still very competitive.

Table 4.10 The mean and standard deviation (in parenthesis) of the experimental results of $f_1, f_3, f_5, f_8 - f_{13} (D = 30)$ and Dixon-Szegö test functions $f_{18} - f_{19}$: The results of JADE with and without archive are calculated based on 50 independent runs. The results of Adaptive LEP and BestLevy are taken from [33, Table III], those of NSDE from [89, Tables 1 – 3] and those of jDE from [22, Table III]

	Gen	JADE w/o archive	JADE with archive	Adaptive LEP*	BestLevy[1]	NSDE[1,2]	jDE[1,3]
f_1	1500	**1.8E-60**	*1.3E-54*	6.32E-04	6.59E-04	7.1E-17	1.1E-28
		(8.4E-60)	*(9.2E-54)*	(7.6E-05)	(6.4E-05)	–	(1.0E-28)
f_3	1500	2.8E-15	**8.5E-22**	0.041850	30.628906	*7.9E-16*	0.090075
		(8.2E-15)	**(3.6E-21)**	(0.059696)	(22.113122)	–	(0.080178)
f_5	1500	*3.2E-01*	5.6E-01	43.40	57.75	**5.9E-28**	3.1E-15
		(1.1E+00)	(1.4E+00)	(31.52)	(41.60)	–	(8.3E-15)
f_8	1500	*4.7e+00*	**2.4E+00**	1100.3	670.6	–	–
		(2.3E+01)	**(1.7E+01)**	(58.2)	(52.2)	–	–
f_9	1500	*1.4E-11*	3.8E-11	5.85	1.3E+01	–	**1.5E-15**
		(1.0E-11)	(2.0E-11)	(2.07)	(2.3E+00)		**(4.8E-15)**
f_{10}	1500	**4.4E-15**	**4.4E-15**	1.9E-02	3.1E-02	1.69E-09	7.7E-15
		(0.0E+00)	**(0.0E+00)**	(1.0E-03)	(2.0E-03)	–	(1.4E-15)
f_{11}	1500	**0.0E+00**	*2.0E-04*	2.4E-02	1.8E-02	5.8E-16	**0**
		(0.0E+00)	*(1.4E-03)*	(2.8E-02)	(1.7E-02)	–	**(0)**
f_{12}	1500	**1.6E-32**	**1.6E-32**	6.0E-06	3.0E-05	5.4E-18	6.6E-30
		(5.5E-48)	**(5.5E-48)**	(1.0E-06)	(4.0E-06)	–	(7.9E-30)
f_{13}	1500	*1.4E-32*	**1.3E-32**	9.8E-05	2.6E-04	6.4E-17	5.0E-29
		(1.1E-47)	**(1.1E-47)**	(1.2E-05)	(3.0E-05)	–	(3.9E-29)

[1]In this chapter, we define f_8 as a shifted version of that in [22], [33], and [89] so that its optimal value is equal to 0. [2]Only the mean results are reported for NSDE in [89]. [3]The results of jDE reported in [22] are slightly different from but consistent with those in Table 4.5 obtained in our independent experiments.

4.4.2 *Benefit of JADE's Components*

We are interested in identifying the benefit of incorporating the two components of JADE (without archive): 'DE/current-to-*p*best' mutation strategy and parameter adaptation. For this purpose, we consider two variants of JADE, rand-JADE and nona-JADE (i.e., DE/current-to-*p*best/1/bin). They differ from JADE without archive *only* in that rand-JADE implements 'DE/rand/1' mutation strategy, while nona-JADE does not adopt parameter adaptation (i.e., instead of updating μ_{CR} and μ_F according to (4.16) and (4.18), we set $\mu_{CR} = \mu_F = 0.5$ in (4.15) and (4.17) throughout the optimization process).

Summarized in Table 4.5 – Table 4.8 are simulation results of JADE, rand-JADE, nona-JADE and DE/rand/1/bin. FESS performance indicates that nona-JADE (and rand-JADE in some cases) converges faster than DE/rand/1/bin. Their advantages stem from the relatively greedy mutation strategy and the adaptive parameter control, respectively. However, both JADE variants suffer from frequent premature convergence for some functions. For example, their success rates are clearly less satisfactory in the optimization of f_3 and f_5. This implies that they are incapable of maintaining sufficiently high diversity of population due to the ill-conditioned Hessian matrices of f_3 and f_5 (indicating a narrow valley) around the optimal point.

JADE achieves significantly better performance compared to these variants in terms of both the convergence rate and the reliability. This indicates a mutually beneficial incorporation of the greedy strategy 'DE/current-to-*p*best' and the parameter adaptation. Indeed, a greedy mutation strategy affects the population's diversity in two opposite directions: it tends to decrease the diversity by moving individuals closer to the few best-so-far solutions, but it is also possible to increase the diversity by speeding the evolutionary search in the progress direction. The reliability of a greedy strategy is usually less satisfactory in classic DE algorithms because the former effect of diversity decrease plays the key role. However, a suitable adaptation is able to further increase the progress rate of 'DE/current-to-*p*best' in the promising direction; thus the latter effect of diversity increase may be able to well balance the former effect. This improves both the convergence rate and the reliability of the algorithm.

4.4.3 Evolution of μ_F and μ_{CR} in JADE

The two control parameters F and CR of a classic DE need to be tuned by trial and error for different optimization functions. In JADE, they are controlled by the adaptive parameters μ_F and μ_{CR} that evolve as the algorithm proceeds. The evolutions of μ_F and μ_{CR} are plotted in Figure 4.4 with mean curves and error bars. The error bars imply a clear evolution trend of μ_F and μ_{CR} over 50 independent runs. For example, μ_F and μ_{CR} changes little in the optimization of spherical function f_1. This is reasonable because the shape of the landscape is the same at all evolution stages. For the ellipsoid function f_3, μ_F and μ_{CR} go to steady states after obvious initial changes (i.e., the population, which is initially distributed in a squared region, gradually adapts to the ellipsoid landscape). This is similar to the case of f_5, where μ_F and μ_{CR} reach steady states after the population falls into the narrow valley of the Rosenbrock function. For f_9, the landscape shows different shapes as the algorithm proceeds and thus μ_F and μ_{CR} evolve to different values accordingly.

These observations are consistent with intuition; i.e., there is no single parameter setting of F or CR that is suitable for various problems (even for linearly transformable problems such as the spherical function f_1 and the ellipsoid function f_3) or for different evolution stages of a single problem (e.g., f_9).

We further investigate the effect of the initial values of μ_F and μ_{CR} on the adaptation process. It is observed that, according to experimental results (not reported due to space limitation), the initial value of μ_F has little effect on the performance of the algorithm, while a moderate or large initial value of μ_{CR} is recommended especially for non-separable functions. In general, an initial setting of $\mu_{CR} = \mu_F = 0.5$ works well for all test functions and can be considered as a standard setting for JADE.

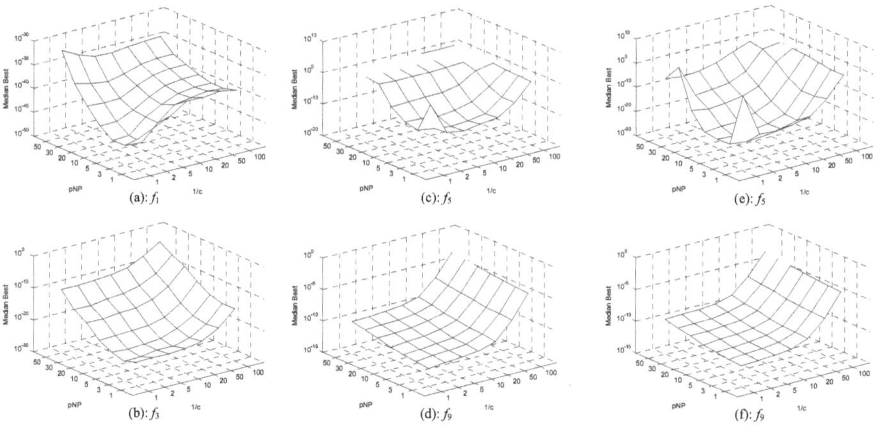

Fig. 4.5 The median of best-so-far function values of f_1, f_3, f_5, and f_9 (D = 30) after 1000, 2000, 2000, 2000 generations, respectively. (a) – (d): JADE without archive; (e) – (f): JADE with archive. Each data point is obtained based on 50 independent runs

4.4.4 Parameter Values of JADE

JADE introduces its own parameters c and p that determine the adaptation rates of μ_{CR} and μ_F and the greediness of the mutation strategy, respectively. Based on their role in JADE, these parameters are thought to be problem insensitive, which is an advantage over the problem-dependent parameter selection (F and CR) in the classic DE. However, it is still interesting to investigate the range where they perform best. As shown in Figure 4.5, the median best values are plotted for JADE with different parameter combinations from $1/c \in \{1, 2, 5, 10, 20, 50, 100\}$ and $p{\cdot}NP \in \{1, 3, 5, 10, 20, \ldots, 50\}$. As expected, a small value of $1/c$ (e.g., $1/c = 1$) or p (e.g., $p{\cdot}NP$ = 1) may lead to less satisfactory results in some cases: The former causes failure of convergence due to the lack of sufficient information to smoothly update CR and F, while the latter is too greedy to maintain the diversity of the population. However, it is clear that JADE is better than or competitive to other algorithms in a large range of c and p (as compared to other algorithms' results in Figure 4.2 and Figure 4.3). In addition, these results also show that JADE (with or without archive) works best with values $1/c \in [5, 20]$ and $p \in [5\%, 20\%]$.

4.5 Performance Analysis in a Noisy Environment

In the previous section, we have mainly investigated the performance of JADE for a set of functions without noise. Although f_7 is a noisy function where JADE shows better performance than other algorithms, it is meaningful to conduct a systematic analysis of JADE for a set of noisy benchmark problems of different characteristics. It is useful for us to understand the potential of JADE/JADE+ as an underlying scheme where explicit noise handling techniques can be implemented. In the literature, researchers have mainly been concerned with two types of noise: (i) noise in the fitness function and (ii) noise in the design variables. In the first case, the noise is additive (and in most cases unbiased) to the fitness function value and the research focus is to reduce the influence of the noise. In the second case, the noise is in the design variables and the main motivation is to find an optimal solution that is insensitive to noise (i.e., search for robust solutions). The optimization in a noisy environment has recently drawn more attention and been developed as an independent research topic [90]. As the theme of this monograph is parameter adaptation, we can evaluate the performance of JADE for optimization problems where noise is additive to the fitness function value. For simplicity, we do not consider such techniques as resampling and thresholding [90], which have been developed in the literature to explicitly address the problem of noise.

Like other studies in the evolutionary optimization field [91], [92], [93], the noisy version of a benchmark function is defined as:

$$f_{\mathrm{n}}(\mathbf{x}) = f(\mathbf{x}) + \mathrm{randn}(0, \sigma^2) \tag{4.21}$$

where $f(\mathbf{x})$ is the clean or noise-free function, $f_{\mathrm{n}}(\mathbf{x})$ is the corresponding noisy function, and randn(0, σ^2) is normally distributed noise with zero mean and standard deviation σ.

Following the settings in [93], we set the population size to be 100 for the optimization of a set of 50-dimensional functions f_1, f_5, f_9, f_{10} and f_{11}. We report in Table 4.11 the final results of each algorithm after 10^5 function evaluations which are averaged over 50 independent runs. For the analysis of the results, we have reported the "true" fitness, i.e., the best noisy fitness value obtained by an algorithm minus the corresponding noise value. Other than the different parameter adaptive DE algorithms, JADE is also compared to the Opposition-Based Differential Evolution (ODE) in [93].

As expected, all algorithms perform much worse in the presence of noise compared to the results for clean functions (i.e., $\sigma^2 = 0$) and the performance gets worse as the magnitude of the noise increases. The optimal function values obtained by different algorithms are of the same order of magnitude as the noise in the case of f_1 and f_{11}. It is however much greater than the magnitude of noise in other cases. Similar to the observations in the literature [91], it seems that the noise confuses the optimization algorithms substantially. It shows that the classic DE is least efficient to optimize the noisy functions. The performance is clearly improved by different parameter adaptive DE algorithms as well as ODE. Among these algorithms, we observe that JADE+ achieves the best results in almost all cases in terms of both the accuracy and the variance of the final results. JADE's performance is usually better than other algorithms but is worse than JADE+.

The inferiority of JADE to JADE+ is reasonable considering that the population of JADE+ is more diversified than that of JADE, which is helpful to increase the "signal"-to-noise ratio and thus reduce the effect of noise. By signal, we mean the difference of the fitness values of the two vectors **x** and **u** in the selection operation (4.14). For a population that is more diversified, the fitness difference tends to be larger. The negative effect of noise is therefore reduced. This is similar to the

Table 4.11 A comparison of the average optimization performance (standard deviation in parenthesis) of different algorithms in a noisy environment

σ^2	Algorithms	f_1	f_5	f_9	f_{10}	f_{11}
0	JADE	**0.00 (0.000)**	*27.920 (10.641)*	*8.861 (1.574)*	**0.00 (0.000)**	**0.00 (0.002)**
	JADE+	**0.00 (0.000)**	**24.162 (8.528)**	**1.103 (0.333)**	**0.00 (0.000)**	**0.00 (0.001)**
	jDE	**0.00 (0.000)**	51.228 (30.035)	14.076 (6.350)	0.007 (0.084)	0.001 (0.004)
	SaDE	**0.00 (0.000)**	37.173 (12.309)	22.344 (14.090)	**0.00 (0.000)**	**0.00 (0.001)**
	SaNSDE	**0.00 (0.000)**	41.055 (12.435)	41.082 (39.871)	**0.00 (0.000)**	**0.00 (0.001)**
	Classic DE	0.003 (0.001)	74.443 (45.372)	371.338 (17.212)	0.014 (0.004)	0.004 (0.002)
	ODE	**0.00 (0.000)**	52.079 (24.807)	142.933 (78.526)	**0.00 (0.000)**	0.001 (0.003)
0.5	JADE	0.894 (0.211)	61.505 (28.735)	*21.172 (1.750)*	*4.733 (2.289)*	1.872 (0.220)
	JADE+	**0.686 (0.160)**	55.676 (20.019)	**14.259 (1.430)**	**3.817 (1.783)**	**1.699 (0.174)**
	jDE	0.981 (0.228)	72.320 (35.727)	25.527 (4.640)	10.612 (8.168)	1.959 (0.237)
	SaDE	1.080 (0.344)	**52.613 (14.655)**	35.516 (14.802)	12.800 (8.514)	2.040 (0.311)
	SaNSDE	0.948 (0.319)	*53.038 (17.716)*	49.164 (33.808)	12.731 (9.051)	1.898 (0.301)
	Classic DE	0.938 (0.201)	81.257 (52.682)	373.966 (17.133)	21.180 (1.488)	1.895 (0.188)
	ODE	*0.874 (0.173)*	56.376 (24.826)	126.863 (61.624)	10.607 (8.679)	*1.844 (0.209)*
1.0	JADE	*1.193 (0.292)*	63.971 (30.935)	*26.883 (2.135)*	**19.588 (2.937)**	*2.198 (0.276)*
	JADE+	**0.968 (0.240)**	**53.605 (17.296)**	**21.198 (1.964)**	*20.665 (2.235)*	**1.970 (0.207)**
	jDE	1.363 (0.327)	74.265 (34.580)	31.310 (4.236)	20.817 (1.435)	2.315 (0.337)
	SaDE	1.542 (0.466)	*55.838 (19.673)*	38.706 (12.160)	20.750 (1.809)	2.441 (0.481)
	SaNSDE	1.279 (0.511)	56.392 (20.189)	52.802 (30.514)	20.736 (2.779)	2.264 (0.490)
	Classic DE	1.830 (0.372)	76.651 (44.276)	372.181 (15.154)	21.617 (0.229)	2.717 (0.378)
	ODE	1.729 (0.484)	62.054 (33.126)	156.033 (66.979)	20.798 (3.102)	2.629 (0.436)

observation in [94] where the authors state that "the signal-to-noise ratio is most likely improved, because the solutions in the population become more diverse."

As the theme here is parameter adaptation, we merely provide simple analysis of the effect of the noise on the performance of a differential evolution algorithm. According to (4.21), we can calculate the probability of making a wrong decision in the selection operation (4.14) as follows,

$$\Pr\{f_n(\mathbf{u}) \geq f_n(\mathbf{x}) \,|\, f(\mathbf{u}) < f(\mathbf{x})\} = \Phi\left(-\frac{|f(\mathbf{u}) - f(\mathbf{x})|}{\sqrt{2}\sigma}\right) \tag{4.22}$$

or

$$\Pr\{f_n(\mathbf{u}) < f_n(\mathbf{x}) \,|\, f(\mathbf{u}) \geq f(\mathbf{x})\} = \Phi\left(-\frac{|f(\mathbf{u}) - f(\mathbf{x})|}{\sqrt{2}\sigma}\right) \tag{4.23}$$

where $\Phi(x)$ denotes the cumulative probability function of a standard normal distribution. Denote SNR as the signal-to-noise ratio, i.e., SNR = $|f(\mathbf{u}) - f(\mathbf{x})|/\sigma$. The probability of a wrong decision in the selection is equal to 24.0% if SNR = 1, 7.9% if SNR = 2, or 1.7% if SNR = 3. Furthermore, we note that

$$\Phi(x) = \frac{1}{\pi}\int_0^{\pi/2} \exp\left(\frac{-x^2}{2\sin^2\theta}\right) d\theta \leq \frac{1}{2}\exp\left(-\frac{x^2}{2}\right), \text{ for } x \leq 0, \tag{4.24}$$

where the first step follows the Craig's identity [95] and the inequality is obtained by setting $\theta = \pi/2$. Thus, from (4.22), (4.23) and (4.24), we have

$$\Pr\{\text{wrong decision in the selection operation}\} \leq \frac{1}{2}\exp\left(-\frac{|f(\mathbf{u})-f(\mathbf{x})|^2}{4\sigma^2}\right), \tag{4.25}$$

which decreases to zero super-exponentially faster when the SNR increases. Thus, it is expected that the noise has little effect on the performance of DE if the SNR is sufficiently large (say SNP ≥ 3); otherwise its effect may not be negligible. Thus, a promising strategy is to utilize the original DE algorithm (classic DE or adaptive DE) in high SNR regions while implementing specific techniques such as resampling and thresholding [90] when a low SNR is detected in the optimization process.

4.6 Scalability Analysis for High-Dimensional Problems

In the literature, the performance of different DE algorithms has mainly been analyzed for a variety of so-called high-dimensional benchmark functions whose dimension is mainly up to 100 [96], [97], [98], [99], [100]. It is meaningful to investigate if and how these algorithms scale well for even higher dimensional problems. In this section, we investigate the scalability of JADE and other different adaptive DE algorithms for the set of classic benchmark functions in Table 4.3 and a new set of test functions proposed in the CEC 2005 Special Session [101], as shown in Table 4.12.

For fair comparison, we follow the parameter settings in the original papers of jDE [22], SaNSDE [25] and SaDE [21], except that the Quasi-Newton local search is disabled in SaDE. The parameters of JADE/JADE+ are set to be $c = 0.1$ and $p = 0.05$ as in the previous sections. The archive size is set to be relatively large $|\mathbf{A}| = 3NP$, as it has been show that the archive is capable of improving the performance especially for high-dimensional problems in Chapter 4.4.

All the simulation results are obtained based on 50 (if $D < 1000$) or 20 (if $D = 1000$) independent runs due to limited computing resources. For clarity, the results

Table 4.12 Some test functions in CEC2005 Special Session [101]

Function	Description
f_{cec1}	Shifted Sphere Function
f_{cec3}	Shifted Rotated High Conditioned Elliptic Function
f_{cec5}	Schwefel Problem 2.6 with Global Optimum on Bounds
f_{cec6}	Shifted Rosenbrock Function
f_{cec8}	Shifted Rotated Ackley Function with Global Optimum on Bounds
f_{cec9}	Shifted Rastrigin Function
f_{cec10}	Shifted Rotated Rastrigin Function
f_{cec13}	Expanded Extended Griewank + Rosenbrock Function

of the best and second algorithms are marked in **boldface** and *italic*, respectively if not all or most algorithms produce identical or very similar results.

4.6.1 Comparison of Different Adaptive DE Algorithms

To investigate the scalability of different algorithms, we consider the two performance metrics introduced in Chapter 4.4: the success rate (SR) and the average number of function evaluations in successful runs (FESS) . For the sake of scalability analysis, we define a successful convergence as follows: An experiment is considered successful if the best solution is found in a sufficiently large number of generations (10^5generations in this section) with sufficient accuracy: 10^{-1} for the noisy function f_7, 10^0 for the step function f_6, or $10^{-5}D$ in other cases. The two performance metrics, FESS and SR, are useful to evaluate an algorithm's convergence rate (in successful runs) and reliability (in all runs), respectively. They are summarized in Table 4.13 and Table 4.14 for problems of dimension from 30 to 1000.

It is clear that JADE+ converges fastest for twelve out of the thirteen functions of dimension from 30 to 1000. Only in the case of f_8, JADE+ converges about 25% slower than jDE but is still faster than or competitive to other algorithms. JADE+ is about 10% faster than JADE on average for 30-D problems and its benefit further increases when the problem dimension increases. The convergence graph of JADE+ is plotted in Figure 4.6 to provide some insights into its convergence behavior.

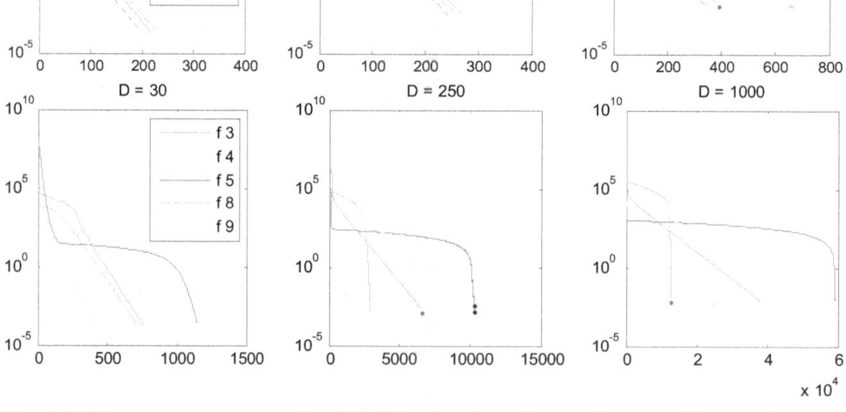

Fig. 4.6 The convergence graph of JADE+ for $f_1 - f_5$ and $f_8 - f_{13}$. The x-axis shows the number of generations. Each curve shows the median of the best-so-far function values obtained at each generation

Table 4.13 Experimental results of different adaptive differential evolution algorithms

D	Algorithm		f_1	f_2	f_3	f_4	f_5	f_6	f_7
30	SaDE	FESS	4.2E+04	5.8E+04	1.1E+05	1.4E+05	2.3E+05	2.3E+04	9.0E+03
		SR	100%	100%	100%	100%	8%	100%	100%
	jDE	FESS	3.6E+04	4.4E+04	2.1E+05	1.5E+05	4.6E+05	2.1E+04	9.6E+03
		SR	100%	100%	100%	100%	98%	100%	100%
	SaNSDE	FESS	4.5E+04	6.5E+04	1.2E+05	1.9E+05	2.2E+05	2.6E+04	1.0E+04
		SR	100%	100%	100%	100%	94%	100%	100%
	JADE	FESS	*2.1E+04*	*3.4E+04*	**5.0E+04**	*4.4E+04*	*1.1E+05*	*1.2E+04*	*4.0E+03*
		SR	100%	100%	100%	100%	92%	100%	100%
	JADE+	FESS	**1.9E+04**	**3.0E+04**	*6.8E+04*	**3.7E+04**	**1.0E+05**	**1.1E+04**	**3.7E+03**
		SR	100%	100%	94%	100%	94%	100%	100%
100	SaDE	FESS	3.3E+05	6.7E+05	1.1E+06	–	–	1.9E+05	8.9E+04
		SR	100%	100%	100%	0%	0%	100%	100%
	jDE	FESS	2.7E+05	3.2E+05	3.4E+06	2.2E+06	6.1E+06	1.6E+05	1.0E+05
		SR	100%	100%	100%	100%	98%	100%	100%
	SaNSDE	FESS	2.7E+05	4.2E+05	1.5E+06	–	2.7E+06	1.8E+05	8.6E+04
		SR	100%	100%	100%	0%	88%	100%	100%
	JADE	FESS	*8.3E+04*	*1.2E+05*	**5.7E+05**	*3.3E+05*	*1.2E+06*	*5.0E+04*	*2.4E+04*
		SR	100%	100%	100%	100%	90%	100%	100%
	JADE+	FESS	**7.4E+04**	**1.1E+05**	*7.1E+05*	**1.8E+05**	**1.1E+06**	**4.4E+04**	**2.2E+04**
		SR	100%	100%	86%	100%	100%	100%	100%
250	SaDE	FESS	1.4E+06	3.8E+06	8.0E+06	–	–	8.0E+05	4.2E+05
		SR	100%	100%	100%	0%	0%	100%	100%
	jDE	FESS	1.2E+06	1.4E+06	2.7E+07	1.9E+07	4.5E+07	7.2E+05	5.3E+05
		SR	100%	100%	100%	100%	100%	100%	100%
	SaNSDE	FESS	1.1E+06	1.6E+06	1.3E+07	–	1.7E+07	1.6E+06	4.0E+05
		SR	100%	100%	100%	0%	80%	100%	100%
	JADE	FESS	*2.5E+05*	*3.8E+05*	**3.7E+06**	*2.5E+06*	*8.8E+06*	*1.8E+05*	*8.0E+04*
		SR	100%	100%	100%	100%	82%	100%	100%
	JADE+	FESS	**2.1E+05**	**2.8E+05**	*5.0E+06*	**1.4E+06**	**7.7E+06**	**1.2E+05**	**7.1E+04**
		SR	100%	100%	98%	100%	100%	100%	100%
500	SaDE	FESS	5.2E+06	9.1E+06	3.5E+07	–	–	2.9E+06	1.6E+06
		SR	100%	100%	100%	0%	0%	100%	100%
	jDE	FESS	3.6E+06	4.4E+06	1.3E+08	1.2E+08	–	2.3E+06	1.9E+06
		SR	100%	100%	100%	100%	0%	100%	100%
	SaNSDE	FESS	3.3E+06	5.2E+06	9.8E+07	–	7.5E+07	1.0E+07	1.3E+06
		SR	100%	100%	100%	0%	80%	100%	100%
	JADE	FESS	*6.8E+05*	*1.3E+06*	**1.6E+07**	*2.3E+07*	*3.9E+07*	*2.2E+06*	*2.0E+05*
		SR	100%	100%	100%	100%	80%	100%	100%
	JADE+	FESS	**4.6E+05**	**6.7E+05**	*2.3E+07*	**1.5E+07**	**3.6E+07**	**2.8E+05**	**1.7E+05**
		SR	100%	100%	100%	100%	100%	100%	100%
1000	SaDE	FESS	2.2E+07	2.7E+07	–	–	–	1.1E+07	7.5E+06
		SR	100%	100%	0%	0%	0%	100%	100%
	jDE	FESS	1.1E+07	1.4E+07	–	–	–	*7.4E+06*	6.4E+06
		SR	100%	100%	0%	0%	0%	100%	100%
	SaNSDE	FESS	1.1E+07	2.2E+07	–	–	–	6.1E+07	4.8E+06
		SR	100%	100%	0%	0%	0%	100%	100%
	JADE	FESS	*2.0E+06*	*1.1E+07*	**7.0E+07**	–	*2.1E+06*	1.7E+07	*5.5E+05*
		SR	100%	50%	100%	0%	80%	100%	100%
	JADE+	FESS	**1.2E+06**	**2.0E+06**	*9.8E+07*	–	**1.8E+08**	**1.1E+06**	**4.3E+05**
		SR	100%	100%	100%	0%	100%	100%	100%

Table 4.14 Experimental results of different adaptive differential evolution algorithms

D	Algorithm		f_8	f_9	f_{10}	f_{11}	f_{12}	f_{13}
30	SaDE	FESS	*6.0E+04*	9.6E+04	5.6E+04	4.7E+04	3.9E+04	4.1E+04
		SR	100%	100%	100%	100%	100%	100%
	jDE	FESS	**4.5E+04**	*8.7E+04*	4.6E+04	3.9E+04	3.2E+04	3.6E+04
		SR	100%	100%	100%	100%	100%	100%
	SaNSDE	FESS	6.3E+04	1.4E+05	5.9E+04	4.9E+04	4.1E+04	4.4E+04
		SR	100%	100%	100%	100%	100%	100%
	JADE	FESS	7.0E+04	9.1E+04	*2.8E+04*	*2.6E+04*	*1.8E+04*	*2.1E+04*
		SR	100%	100%	100%	100%	100%	100%
	JADE+	FESS	6.3E+04	**7.9E+04**	**2.6E+04**	**2.0E+04**	**1.6E+04**	**1.8E+04**
		SR	100%	100%	100%	100%	100%	100%
100	SaDE	FESS	6.9E+05	9.7E+05	4.1E+05	3.2E+05	3.5E+05	3.6E+05
		SR	100%	100%	100%	100%	100%	100%
	jDE	FESS	**3.8E+05**	*8.7E+05*	3.0E+05	2.6E+05	2.5E+05	2.8E+05
		SR	100%	100%	100%	100%	100%	100%
	SaNSDE	FESS	5.8E+05	1.7E+06	3.2E+05	2.7E+05	2.4E+05	2.8E+05
		SR	100%	100%	100%	92%	100%	100%
	JADE	FESS	6.4E+05	1.0E+06	*9.6E+04*	*8.3E+04*	*6.2E+04*	*7.8E+04*
		SR	100%	100%	100%	100%	100%	100%
	JADE+	FESS	*5.5E+05*	**8.4E+05**	**8.6E+04**	**7.3E+04**	**5.4E+04**	**6.8E+04**
		SR	100%	100%	100%	100%	100%	100%
250	SaDE	FESS	4.0E+06	5.6E+06	1.6E+06	1.3E+06	1.8E+06	2.1E+06
		SR	100%	100%	100%	96%	100%	100%
	jDE	FESS	**2.0E+06**	*5.1E+06*	1.2E+06	1.1E+06	1.2E+06	1.3E+06
		SR	100%	100%	100%	100%	100%	100%
	SaNSDE	FESS	3.2E+06	8.9E+06	1.4E+06	1.0E+06	1.5E+06	1.7E+06
		SR	100%	100%	12%	84%	62%	80%
	JADE	FESS	3.6E+06	6.4E+06	*2.8E+05*	*2.5E+05*	*1.8E+05*	*2.3E+05*
		SR	100%	100%	100%	100%	92%	98%
	JADE+	FESS	*2.2E+06*	**4.7E+06**	**2.2E+05**	**2.0E+05**	**1.4E+05**	**1.8E+05**
		SR	100%	100%	100%	100%	100%	100%
500	SaDE	FESS	1.4E+07	2.0E+07	5.7E+06	4.3E+06	8.7E+06	9.5E+06
		SR	100%	100%	100%	100%	100%	98%
	jDE	FESS	**6.9E+06**	**1.9E+07**	3.5E+06	3.2E+06	4.3E+06	4.4E+06
		SR	100%	100%	100%	100%	100%	100%
	SaNSDE	FESS	1.5E+07	2.9E+07	–	3.0E+06	1.1E+07	1.0E+07
		SR	100%	68%	0%	88%	32%	88%
	JADE	FESS	1.6E+07	2.7E+07	*7.5E+05*	*6.5E+05*	*4.4E+05*	*5.9E+05*
		SR	100%	100%	100%	100%	88%	80%
	JADE+	FESS	*8.9E+06*	**1.9E+07**	**4.6E+05**	**4.4E+05**	**2.9E+05**	**4.2E+05**
		SR	100%	100%	100%	100%	100%	94%
1000	SaDE	FESS	*3.5E+07*	**7.1E+07**	–	1.4E+07	4.4E+07	5.4E+07
		SR	100%	100%	0%	100%	95%	100%
	jDE	FESS	**2.5E+07**	*7.4E+07*	1.0E+07	9.9E+06	1.5E+07	1.5E+07
		SR	100%	100%	100%	100%	100%	100%
	SaNSDE	FESS	5.6E+07	–	–	9.9E+06	4.9E+07	1.7E+08
		SR	100%	0%	0%	100%	75%	75%
	JADE	FESS	6.9E+07	1.1E+08	*3.2E+06*	*1.9E+06*	*1.2E+06*	*1.7E+06*
		SR	100%	100%	65%	100%	100%	75%
	JADE+	FESS	3.8E+07	*7.4E+07*	**1.1E+06**	**1.1E+06**	**7.2E+05**	**1.0E+06**
		SR	100%	100%	100%	100%	100%	90%

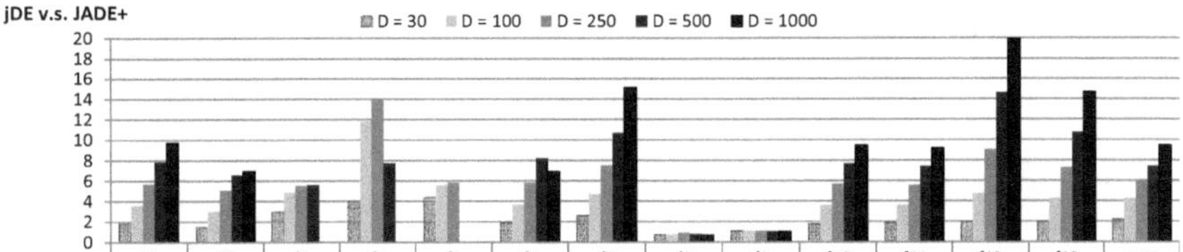

	f1	f2	f3	f4	f5	f6	f7	f8	f9	f10	f11	f12	f13	ave
D = 30	1.89	1.48	2.98	4.01	4.37	1.95	2.60	0.73	1.11	1.78	1.91	2.02	2.01	2.22
D = 100	3.56	2.99	4.83	11.74	5.58	3.61	4.66	0.70	1.05	3.54	3.53	4.68	4.20	4.20
D = 250	5.68	5.05	5.49	14.04	5.78	5.82	7.46	0.88	1.07	5.66	5.53	9.04	7.26	6.06
D = 500	7.82	6.55	5.56	7.70	0.00	8.18	10.68	0.78	1.03	7.66	7.37	14.60	10.70	7.39
D = 1000	9.70	6.94	0.00	0.00	0.00	6.89	15.12	0.66	1.00	9.45	9.17	20.86	14.66	9.44

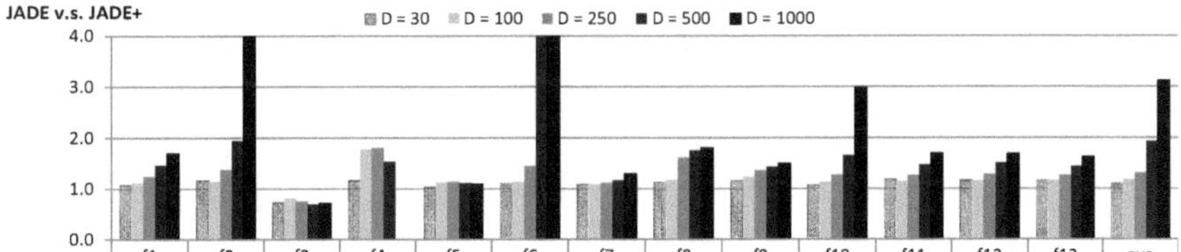

	f1	f2	f3	f4	f5	f6	f7	f8	f9	f10	f11	f12	f13	ave
D = 30	1.08	1.17	0.73	1.17	1.04	1.11	1.09	1.13	1.16	1.08	1.18	1.17	1.16	1.10
D = 100	1.11	1.14	0.81	1.77	1.12	1.13	1.09	1.17	1.22	1.12	1.13	1.15	1.16	1.16
D = 250	1.24	1.37	0.75	1.80	1.14	1.45	1.12	1.61	1.36	1.27	1.26	1.28	1.27	1.30
D = 500	1.46	1.95	0.69	1.53	1.11	7.83	1.17	1.75	1.42	1.66	1.47	1.50	1.43	1.92
D = 1000	1.70	5.80	0.71	0.00	1.09	15.51	1.29	1.80	1.49	2.98	1.70	1.68	1.62	3.11

Fig. 4.7 The speedup ratio of JADE+ with respect to jDE and JADE. The last column "ave" of each table is the value averaged over all test functions. Note that some results are not available when the success rate of an algorithm is 0

To show the superiority of JADE+, we further calculate the ratio, termed the *speed-up ratio* , of the FESS of an algorithm to that of JADE+. The ratio indicates how JADE+ performs relative to other algorithms. It is clear from Figure 4.7 and Figure 4.8 that the benefit of JADE+ over other algorithms usually gets more prominent as the problem dimension increases. The average speedup ratio ranges from 1.10 to 3.12 (in the case of JADE), 2.22 to 9.44 (jDE), 2.14 to 21.20 (SaDE), or 2.41 to 41.52 (SaNSDE) as the problem dimension rises from 30 to 1000.

As far as the speedup ratio for a single function is concerned, we can observe a clearly increasing trend of speedup ratios as problem dimension increases in the case of f_1, f_2, f_6, f_7, and $f_{10} - f_{11}$. In other cases, the speed-up ratio is roughly a constant for different problem dimensions (e.g., f_8 and f_9) or slightly decreases but is still greater than 1. For example, the speedup ratio of JADE+ with respect to jDE is about 0.7 or 1.0 in the case of f_8 or f_9, respectively. Note that f_8 is the only function where the speedup ratio is always less than 1. In this case, JADE+'s performance is slightly worse than jDE but is still better than or competitive to other algorithms.

It is also important to compare the reliability of different algorithms, as shown in Table 4.13 and Table 4.14. As expected, jDE works very reliably because it adopts the 'DE/rand/1' mutation strategy which is less greedy and more robust than other strategies. Its success rate is 100% in most cases, except for the 30- and 100-D

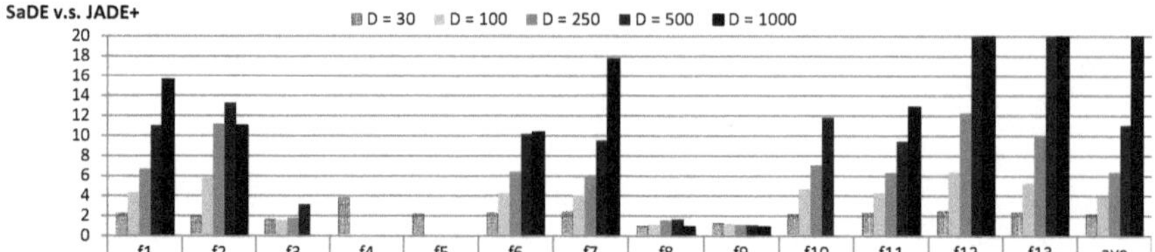

	f1	f2	f3	f4	f5	f6	f7	f8	f9	f10	f11	f12	f13	ave
D = 30	2.17	1.96	1.62	3.75	2.17	2.22	2.43	0.96	1.25	2.16	2.29	2.45	2.33	2.13
D = 100	4.35	5.79	1.60	0.00	0.00	4.27	4.07	1.15	1.24	4.70	4.30	6.35	5.28	3.92
D = 250	6.67	11.18	1.79	0.00	0.00	6.40	5.90	1.53	1.13	7.03	6.30	12.25	9.98	6.38
D = 500	10.98	13.24	3.13	0.00	0.00	10.16	9.51	1.62	1.06	11.83	9.40	28.81	21.58	11.03
D = 1000	15.60	10.99	0.00	0.00	0.00	10.36	17.73	0.92	0.90	0.00	12.88	60.60	54.01	20.44

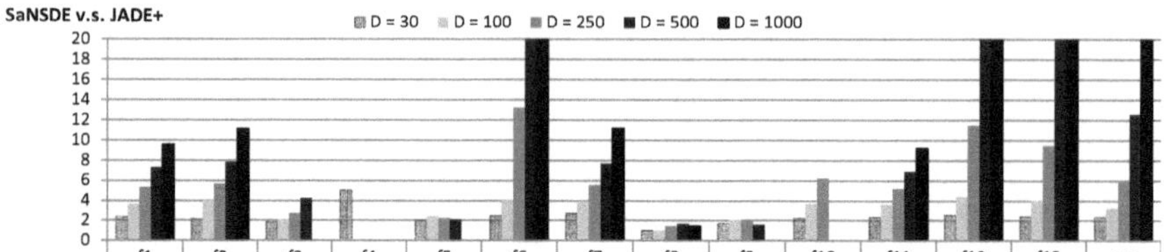

	f1	f2	f3	f4	f5	f6	f7	f8	f9	f10	f11	f12	f13	ave
D = 30	2.34	2.15	1.84	5.06	2.06	2.49	2.77	1.02	1.75	2.28	2.40	2.60	2.48	2.40
D = 100	3.66	3.96	2.12	0.00	2.43	4.02	3.94	1.06	2.08	3.70	3.65	4.44	4.11	3.26
D = 250	5.31	5.65	2.71	0.00	2.24	13.20	5.54	1.45	1.90	6.21	5.19	11.46	9.45	5.86
D = 500	7.26	7.82	4.21	0.00	2.10	35.05	7.66	1.68	1.56	0.00	6.88	38.63	25.03	12.53
D = 1000	9.56	11.12	0.00	0.00	0.00	51.76	11.16	1.45	0.00	0.00	9.20	66.36	168.45	41.13

Fig. 4.8 The speedup ratio of JADE+ with respect to SaDE and SaNSDE. The last column "ave" of each table is the value averaged over all test functions. Note that some results are not available when the success rate of an algorithm is 0

Rosenbrock function f_5 (mainly due to premature convergence), the 1000-Df_3 and f_4, and the 500- and 1000-D f_5 (mainly due to slow convergence). The success rate of JADE+ is equal to 100% in most cases, except for the 30-D f_5, the 30- to 250-D f_3, and the 500- and 1000-D f_{13}. Its reliability is slightly inferior yet very competitive to jDE. It is interesting to observe that the reliability (success rate) of an algorithm is not necessarily degraded when the problem dimension increases. Take the optimization of f_5 as an example. JADE+ causes false convergence only when the problem dimension is low, although the success rate of JADE and SaNSDE decreases when the problem dimension is high.

The reliability of JADE is inferior to that of JADE+ and jDE, as it may cause false convergence in the case of f_5 ($D = 30$ to 1000), 1000-D f_2, f_4 and f_{10}, 250- and 500-Df_{12}, and 250- to 1000-D f_{13}. SaDE has difficulty to optimize f_4 for $D \geq$ 100 and f_5 for $D = 30$ to 1000 and causes false convergence in the case of 1000-Df_3, $f_{10}$and f_{12}, and 500-Df_{13}. The reliability of SaNSDE is competitive to other algorithms for low-dimensional problems ($D = 30$ or 100). However, it causes frequent false convergence in the case of f_4, f_5, $f_9 - f_{13}$ for moderate to high dimensional problems. In addition, we notice that SaNSDE causes frequent false convergence for f_{12} and f_{13} when $D = 250$ and 500, but it seems less difficult to optimize f_{12} and f_{13} when $D = 30$, 100 or 1000. In fact, this is because SaNSDE frequently converges to local minimal values that are slightly less than 0.01 when $D \geq 250$. It is considered as false convergence when $D = 250$ and 500, but it is not when $D = 1000$ and thus the target accuracy is $10^{-5} \cdot D = 0.01$.

4.6.2 Scalability of Different Adaptive DE Algorithms

We next investigate the scalability of different algorithms by considering the FESS as a function of problem dimension D. As a first step, we estimate if FESS can be approximated as a polynomial or super-polynomial (say, exponential) function of D. In Figure 4.9, we consider JADE+ as an example and plot the FESS, sqrt(FESS), and $\log_{10}$(FESS) as functions of D. It is clear that the FESS increases less-exponentially fast and can be well approximated as a polynomial function of D: the order of the polynomial is around 2 for all the 13 benchmark functions. Although not reported in Figure 4.9, the FESS of other algorithms can also be well approximated as a polynomial function.

For the sake of a qualitative analysis, we model the FESS(i, D) of each function i as a polynomial function of D as follows:

$$\mathrm{FESS}(i,D) \approx a(i)D^{b(i)} + c(i), \tag{4.26}$$

where $a(i)$, $b(i)$ and $c(i)$ are three constants which may vary for different optimization problems, or

$$\mathrm{FESS}(i,D) \approx a(i)D^{b} + c(i), \tag{4.27}$$

where b is considered as a constant suitable for all optimization problems. For each algorithm, we define $b(i)$ as its *problem-specific scalability metric* for function i,

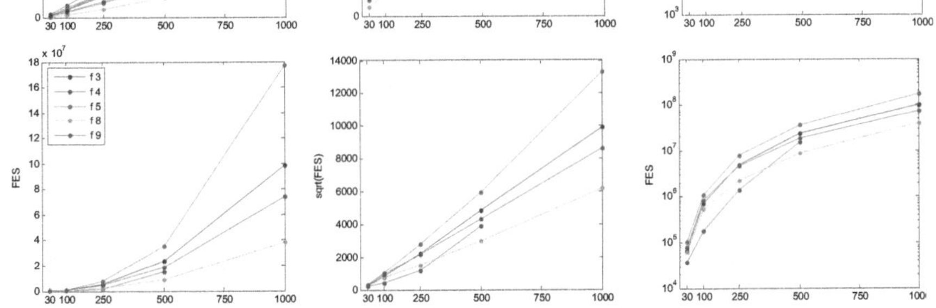

Fig. 4.9 An illustration of FESS as a function of problem dimension D (take JADE+ as an example). The x-axis represents the problem dimension D. From left to right, the three plots show the relationship between FESS, sqrt(FESS), or $\log_{10}$(FESS) and problem dimension D

and b as its *overall scalability metric* for all the functions of interest. The values of $b(i)$ and b can be estimated by minimizing the discrepancy between the approximate FESS and actual FESS. To be specific, we calculate b and $b(i)$ by solving two discrepancy minimization problems

$$\min_{b,a(i),c(i)} \sum_{i} \sum_{D} \left(1 - \frac{\text{FESS}(i,D)}{a(i)D^{b} + c(i)}\right)^{2}, \tag{4.28}$$

and

$$\min_{b(i),a(i),c(i)} \sum_{D} \left(1 - \frac{\text{FESS}(i,D)}{a(i)D^{b(i)} + c(i)}\right)^{2}, \tag{4.29}$$

which, for the sake of solving $a(i)$, $b(i)$ and $c(i)$, is applicable only when there are at least three finite FESS values. Based on the FESS values reported in Table 4.13 and Table 4.14, we solve the two optimization problems in (4.28) and (4.29) for each adaptive DE algorithm. The overall scalability metric in (4.29) is calculated by taking into account all the 13 functions or 11 functions (except f_4 and f_5 for which the problem-specific scalability cannot be calculated in the case of SaDE and SaNSDE). The calculated scalability metrics are summarized in Table 4.15. It is clear that JADE+ shows the best scalability among the different algorithms: its overall scalability metric achieves the best value (less than 1.7), and its problem-specific scalability metric is the best for 8 out of the 13 functions: f_1, f_2, f_6, f_7, $f_{10} - f_{13}$ and is competitive in other cases. jDE show the second best overall scalability when all the 13 functions are concerned, while JADE is the second best for the 11 functions excluding f_4 and f_5. The overall scalability metrics of SaDE and SaNSDE are less satisfactory compared to other algorithms.

Table 4.15 A comparison of the problem-specific and overall scalability metric of different algorithms

	f_1	f_2	f_3	f_4	f_5	f_6	f_7	f_8	f_9	f_{10}	f_{11}	f_{12}	f_{13}	overall*	
SaDE	1.84	1.52	2.14	–	–	1.78	1.93	**1.71**	1.86	1.66	1.67	2.14	2.21	–	(1.876)
jDE	1.63	1.64	2.26	**2.48**	2.17	1.66	1.79	1.84	1.93	1.53	1.58	1.78	1.73	1.904	(1.842)
SaNSDE	1.64	1.75	2.63	**–**	**2.07**	2.55	1.74	2.02	**1.74**	1.63	1.59	2.44	2.89	–	(1.993)
JADE	1.44	2.18	**2.10**	2.71	2.25	2.69	1.35	2.05	2.04	1.65	1.44	1.36	1.38	1.99	(1.838)
JADE+	**1.22**	**1.34**	2.16	2.88	2.22	**1.43**	**1.27**	1.87	1.95	**1.15**	**1.20**	**1.16**	**1.21**	**1.697**	**(1.613)**

* The overall scalability metrics in the second to last column are calculated based on all the 13 functions, while those in the last column (reported in the parenthesis) based on 11 functions except f_4 and f_5.

4.6.3 Comparison of JADE+ with Coevolutionary Algorithms

As stated in the Introduction, parameter adaptive DE algorithms can be used as the underlying scheme where new techniques such as the cooperative coevolution (CC) can be incorporated. It is still interesting, though, to compare a parameter adaptive differential evolution algorithm with the existing DE-based cooperative coevolution algorithm in the literature. For this purpose, we compare JADE+ to two recently proposed CC algorithms, DECC-O [102] and DECC-G [103] , which have shown their advantage over earlier CC algorithms such as an evolutionary programming based FEPCC [104]. We also include the results of SaNSDE (reported in [103]), as it is the underlying scheme of DECC-G.

As shown in Table 4.16, all the results are obtained based on twenty-five independent runs, among which the results of CC algorithms come from [103]. It is of no surprise that JADE+ achieves better solutions than SaNSDE in seven out of eight CEC 2005 test functions while their performance is the same in the case of f_{cec8}. However, it is interesting to observe that JADE+ also performs significantly better than both DECC-O and DECC-G in five out of the eight functions in both cases of $D = 500$ and $D = 1000$. Among these functions, f_{cec3} and f_{cec10} are rotated problems and their variables are highly correlated in calculating the objective function value. It clearly shows that CC algorithms fail to work efficiently for problems whose variables are strongly correlated. JADE+ also outperforms the two CC algorithms in the case of the shifted sphere function f_{cec1}, which is separable, and the shifted Rosenbrock function f_{cec5}, of which only the adjacent variables are directly correlated. On the contrary, the two CC algorithms achieve better solutions than JADE+ for the separable function f_{cec9} and another function f_{cec13}, of which only the adjacent variables are directly correlated. f_{cec8} is a shifted rotated function with global optimum on bounds. A closer look reveals that all the algorithms achieve very similar results, around 2.16E+1, which are far away from the optimal value 0.

Table 4.16 A comparison of JADE+ to cooperative coevolutionary algorithms

D	Algorithm	f_{cec1}	f_{cec3}	f_{cec5}	f_{cec6}	f_{cec8}	f_{cec9}	f_{cec10}	f_{cec13}
500	SaNSDE	2.61E-11	6.88E+8	4.96E+5	2.71E+3	*2.15E+1*	6.60E+2	6.97E+3	2.53E+2
	DECC-O	1.04E-12	4.78E+8	2.40E+5	1.71E+3	**2.14E+1**	**8.66E+0**	1.50E+4	**2.81E+1**
	DECC-G	*3.71E-13*	*3.06E+8*	*1.15E+5*	*1.56E+3*	2.16E+1	*4.50E+2*	*5.33E+3*	*2.09E+2*
	JADE+	**0.00E+0**	**3.11E+7**	**2.50E+4**	**5.27E+2**	2.16E+1	2.31E+3	**2.62E+2**	2.11E+2
1000	SaNSDE	1.17E+0	2.34E+9	*5.03E+5*	1.35E+4	*2.16E+1*	3.20E+3	1.27E+4	6.61E+2
	DECC-O	3.66E-08	1.08E+9	3.73E+5	3.13E+3	**2.14E+1**	**8.96E+1**	3.18E+4	**7.52E+1**
	DECC-G	*6.84E-13*	*8.11E+8*	2.20E+5	*2.22E+3*	*2.16E+1*	*6.32E+2*	*9.73E+3*	*3.56E+2*
	JADE+	**2.43E-20**	**1.12E+8**	**5.74E+4**	**9.71E+2**	*2.16E+1*	6.52E+3	**5.39E+2**	5.57E+2

According to their reported results, the cooperative coevolutionary algorithms such as FEPCC, DECC-O and DECC-G usually achieve similar or slightly inferior solutions, if the number of function evaluations is increased proportionally for high dimensional problems. The scalability metric of these algorithms is thus close to or slightly greater than 1. It is expected that the performance of JADE+ can be exceeded by CC-based algorithms (for separable or moderately separable problems) if the problem dimension is sufficiently high.[2] This motivates an incorporation of cooperative cooperation techniques to JADE+ to further produce better results.

4.7 Summary

Adaptive parameter control is beneficial to improve the reliability of an evolutionary algorithm. It is natural to incorporate parameter adaptation with a greedy mutation strategy such as the archive-assisted 'DE/current-to-*p*best/1' or 'DE/rand-to-*p*best/1' to increase the convergence rate while maintaining the reliability of the algorithm at a high level. This motivates our consideration of JADE: a parameter adaptive archive-assisted 'DE/current-to-*p*best' or 'DE/rand-to-*p*best' algorithm. In JADE, the mutation strategy utilizes the information of multiple high-quality solutions to improve the population diversity and thus moderate its greediness. The adaptation of JADE is implemented by evolving the mutation factors and crossover probabilities based on their historical success information. We also introduce an external archive to store recently explored inferior solutions and their difference from the current population provides the information about the promising direction towards the optimum.

JADE has been tested on a set of commonly used classic benchmark functions taken from the literature. It shows better or at least competitive optimization performance in terms of the convergence rate and the reliability, when compared to other classic and adaptive DE algorithms, the canonical PSO algorithm and other evolutionary algorithms from the literature.

We have further investigated the scalability of JADE/JADE+ and other parameter adaptive DE algorithms for high dimensional problems. It shows that the number of function evaluations in a successful run of these algorithms can be well approximated as a polynomial function of the problem dimension. The order of the polynomial function is defined as the scalability metric of an algorithm. It shows that the overall scalability metric of JADE+ is equal to 1.697 for the set of 13 benchmark functions, which is clearly better than other competitive algorithms whose overall scalability metric range from 1.8 to 2.0. The average convergence speedup of JADE+ with respect to other algorithms increases from 1.10 to 3.12 (in the case of JADE), 2.22 to 9.44 (jDE), 2.14 to 21.20 (SaDE), or 2.41 to 41.52 (SaNSDE) as the problem dimension increases from 30 to 1000. JADE+ is also compared to two recent DE-based cooperative coevolutionary algorithms. It shows that JADE+ performs best in five out of the eight CEC 2005 benchmark functions, including the

[2] Consider the sphere function f_1 as an example. DECC-G reaches an accuracy of 6.33E-27 for $D = 500$ after 2.50E6 function evaluations [103, Table 2]. Simulation results show that it takes 1.67E6 function evaluations for JADE+ to reach the same accuracy; i.e., JADE+ is 1.50 faster than DECC-G for the 500-dimensional f_1. Considering that the scalability metrics of JADE+ and DECC-G are 1.223 and about 1, respectively, the performance of DECC-G can exceed that of JADE+ when the dimension of f_1 is greater than 3000.

shifted sphere and Rosenbrock functions and some rotated functions. It indicates a potential of better performance by applying JADE+ as the underlying strategy where cooperative coevolution techniques can be implemented.

4.8 Appendix: Stochastic Properties of the Mutation and Crossover of JADE

In this section, we analyze the stochastic properties of the mutation and crossover in a simplified model of JADE. For this purpose, we assume that (i) the selection operation is ignored by simply setting the offspring vector generated by mutation and crossover to be the parent vector in the next generation (it is equivalent to assume that the fitness landscape is a plane), (ii) the parameter adaptation is disabled by setting μ_F and μ_{CR} as constants throughout the optimization process (the F are CR values, however, are still randomly generated).

We do not include the selection and parameter adaptation operations in the analysis because of the complexity caused by their problem-dependent property. The stochastic properties of the mutation and crossover, however, still provide us some insights into the potential performance of JADE. For example, it is useful for us to understand the difference between the new mutation strategy proposed in JADE and that in the classic DE; it helps us understand whether the performance may be degraded if μ_F and μ_{CR} values are set inappropriately; and it provides some idea of how the population diversity changes in lack of selection pressure.

4.8.1 A Simplified Model

Assume the initial population $\{\mathbf{x}_{i,0}= (x_{1,i,0}, x_{2,i,0}, \ldots, x_{D,i,0})|i = 1, 2, \ldots, NP\}$ is randomly generated according to a certain probability distribution (e.g., uniform or normal distribution).

The mutation vector $\mathbf{v}_{i,g}$ is generated according to (4.12), i.e.,

$$\mathbf{v}_{i,g} = \mathbf{x}_{r0,g} + F_i(\mathbf{x}^p_{\text{best},g} - \mathbf{x}_{r0,g}) + F_i(\mathbf{x}_{r1,g} - \mathbf{x}_{r2,g}), \tag{4.30}$$

where $r_0 \neq r_1 \neq r_2 \neq i$ are distinct integers uniformly chosen from the set $\{1, 2, \ldots, NP\}$, and F_i is the mutation factor that is generated according to a Cauchy distribution (4.17)

$$F_i = \text{randc}_i(\mu_F, 0.1), \tag{4.31}$$

and then is truncated to be 1 if $F_i \geq 1$ or regenerated if $F_i \leq 0$. The variance of this truncated distribution is a constant when there is no parameter adaptation (i.e., μ_F is a constant). $\mathbf{x}^p_{\text{best},g}$ is randomly chosen as one of the top $100p\%$ individuals in the current population with $p \in (0, 1]$. If the landscape is a plane or the fitness function is constant, all the individual vectors are equally good and thus, similar to $\mathbf{x}_{r0}$, $\mathbf{x}_{r1}$ and $\mathbf{x}_{r2}$, $\mathbf{x}^p_{\text{best},g}$ is a vector that is randomly chosen from the population. For simplicity, assume $\mathbf{x}^p_{\text{best},g}$ is chosen to be distinct from $\mathbf{x}_{i,g}$. Thus, $\mathbf{x}_{r0}$, $\mathbf{x}_{r1}$, $\mathbf{x}_{r2}$, and $\mathbf{x}^p_{\text{best},g}$ are all independent of $\mathbf{x}_i$.

The offspring vector $\mathbf{u}_{i,g}$ is then generated via the crossover operation (4.13). For the simplicity of theoretical analysis, we simplify the operation in (4.13) as follows:

$$u_{j,i,g} = \begin{cases} v_{j,i,g}, & \text{if } \text{rand}_j(0,1) \le CR_i \\ x_{j,i,g}, & \text{otherwise}, \end{cases} \tag{4.32}$$

where $\text{rand}_j(0, 1)$ is a uniform random number on the interval [0, 1] and newly generated for each j, and the crossover probability CR_i is generated according to a normal distribution (4.15),

$$CR_i = \text{randn}_i(\mu_{CR}, 0.1), \tag{4.33}$$

that is truncated to [0, 1]. The variance of this truncated distribution is a constant when there is no parameter adaptation (i.e., μ_{CR} is a constant). In the original formula of (4.4), we have $u_{j,i,g} = v_{j,i,g}$ if $\text{rand}_j(0,1) \le CR_i$ or $j = j_{\text{rand}}$, where j_{rand} is an integer randomly chosen from 1 to D. The usage of j_{rand} is to avoid the trivial situation of $\mathbf{v}_{i,g} = \mathbf{x}_{i,g}$ by ensuring that at least one component of $\mathbf{v}_{i,g}$ comes from the mutation vector $\mathbf{u}_{i,g}$. The simplification of (4.32) has little effect on the accuracy of the analytical results (especially when $CR_i \cdot D$ is relatively large) but can significantly simplify the derivation.

When the selection operation is ignored, we simply set the offspring vector as the new parent vector in the next generation; i.e.,

$$\mathbf{x}_{i,g+1} = \mathbf{u}_{i,g}. \tag{4.34}$$

For the sake of notational simplicity, we will omit the subscript g and/or $i$when no confusion is possible. In addition, we may use $\mathbf{z}^2$ to denote the product $\mathbf{z}\mathbf{z}'$ of a column vector $\mathbf{z}$ and its transpose.

4.8.2 Mathematical Analysis

We first present main analytical results and then provide a detailed proof.

Proposition 4.1. *If the population $\{\mathbf{x}_i\}$ follows an arbitrary distribution of mean $E(\mathbf{x})$ and variance $Var(\mathbf{x})$, then the means and variances of the mutation vectors $\{\mathbf{v}_i\}$ and offspring vectors $\{\mathbf{u}_i\}$ are given by*

$$E(\mathbf{v}) = E(\mathbf{x}), \tag{4.35}$$

$$Var(\mathbf{v}) = E(1 - 2F + 4F^2)Var(\mathbf{x}), \tag{4.36}$$

$$E(\mathbf{u}) = E(\mathbf{x}), \tag{4.37}$$

$$Var(\mathbf{u}) = < 1 - 2E(CR) + E(2 - 2F + 4F^2)\Phi, Var(\mathbf{x}) >, \tag{4.38}$$

where <A, B> denotes the element-wise product of two matrices A and B, and $\Phi = \{\phi_{j,k}\}$ is a matrix of constants $\phi_{k,k} = E(CR)$ and $\phi_{j,k} = E(CR^2)$, for $j \ne k$.

According to (4.38), the variance $Var(u)$ of each element of $\mathbf{u}$ (i.e., the diagonal term of $Var(\mathbf{u})$) is given as follows:

$$Var(u) = \left[1 + 2E(2F^2 - F) \cdot E(CR)\right] Var(x), \tag{4.39}$$

where $Var(x)$ denotes the variance of each element of $\mathbf{x}$. In general, it is desired that the variation operators, mutation and crossover, can improve the population diversity, since the selection operation tends to reduce it. This is for the sake of a good balance between the capabilities of exploration and exploitation in the optimization search. Consider that $E(CR) > 0$; we require $E(2F^2 - F) > 0$ so that $Var(u) > Var(x)$. Given that the mutation factor F is generated according to the truncated Cauchy distribution in (4.17), the requirement is satisfied when

$$\mu_F < 0.06 \text{ or } \mu_F > 0.32. \tag{4.40}$$

In JADE, we place more weights on larger successful mutation values in the μ_F adaptation of (4.18) and (4.19). According to simulation, the μ_F value rarely becomes less than 0.32 in the evolutionary optimization process of a variety of benchmark functions.

4.8.3 Proof of Proposition 4.1

We need the following lemma in the proof of the proposition.

Lemma 4.1. *Suppose F is a random variable and $\mathbf{z}_1$ and $\mathbf{z}_2$ are two random vectors of the same length. Suppose F, $\mathbf{z}_1$, and $\mathbf{z}_2$ are independent. The variance of $F\mathbf{z}_1 + (1-F)\mathbf{z}_2$ is given by*

$$Var[F\mathbf{z}_1 + (1-F)\mathbf{z}_2] = E(F^2)Var(\mathbf{z}_1) + E[(1-F)^2]Var(\mathbf{z}_2) + Var(F)[E(\mathbf{z}_1 - \mathbf{z}_2)]^2. \tag{4.41}$$

Proof. According to the definition of the variance, we have

$$Var(F\mathbf{z}_1) = E(F^2\mathbf{z}_1^2) - E(F\mathbf{z}_1)^2. \tag{4.42}$$

F and $\mathbf{z}_1$ are independent, so are their squares F^2 and $\mathbf{z}_1^2$. Thus, we have $E(F\mathbf{z}_1) = E(F)E(\mathbf{z}_1)$ and $E(F^2\mathbf{z}_1^2) = E(F^2)E(\mathbf{z}_1^2)$. Eq. (4.42) can be rewritten as follows

$$\begin{aligned} Var(F\mathbf{z}_1) &= E(F^2)E(\mathbf{z}_1^2) - E(F)^2E(\mathbf{z}_1)^2 \\ &= E(F^2)Var(\mathbf{z}_1) + E(F^2)E(\mathbf{z}_1)^2 - E(F)^2E(\mathbf{z}_1)^2 \\ &= E(F^2)Var(\mathbf{z}_1) + Var(F) \cdot E(\mathbf{z}_1)^2. \end{aligned} \tag{4.43}$$

Similarly, we have

$$\begin{aligned} Var[(1-F)\mathbf{z}_2] &= E[(1-F)^2]Var(\mathbf{z}_2) + Var(1-F)E(\mathbf{z}_1)^2 \\ &= E[(1-F)^2]Var(\mathbf{z}_2) + Var(F)E(\mathbf{z}_1)^2. \end{aligned} \tag{4.44}$$

We next calculate the covariance between $F\mathbf{z}_1$ and $(1-F)\mathbf{z}_2$ as follows:

$$\begin{aligned}
&\mathrm{Cov}(F\mathbf{z}_1,(1-F)\mathbf{z}_2)\\
&= E[F(1-F)\mathbf{z}_1\mathbf{z}'_2] - E(F\mathbf{z}_1)E[(1-F)\mathbf{z}'_2]\\
&= E[F(1-F)]E(\mathbf{z}_1)E(\mathbf{z}'_2) - E(F)E(1-F)E(\mathbf{z}_1)E(\mathbf{z}'_2)\\
&= \mathrm{Var}(F)E(\mathbf{z}_1)E(\mathbf{z}'_2).
\end{aligned} \tag{4.45}$$

Similarly, we have

$$\mathrm{Cov}((1-F)\mathbf{z}_2, F\mathbf{z}_1) = \mathrm{Var}(F)E(\mathbf{z}_2)E(\mathbf{z}'_1). \tag{4.46}$$

Adding up the left hand sides of (4.43) – (4.46), we obtained the variance of $F\mathbf{z}_1 + (1-F)\mathbf{z}_2$. Now, (4.41) is readily available after simple manipulation.

Proof (Proposition 4.1). We note that $\mathbf{x}_{r0}, \mathbf{x}_{r1}$, $\mathbf{x}_{r2}$ and $\mathbf{x}^p_{\text{best}}$ are all independently drawn from the current population and are independent of $\mathbf{x}_i$ in the simplified model. They are identically and independently distributed with mean $E(\mathbf{x})$ and variance $Var(\mathbf{x})$. In addition, they are all independent of F_i. According to (4.30), it is easy to have

$$E(\mathbf{v}) = E(\mathbf{x}_{r0}) + E(F_i)\left[E(\mathbf{x}^p_{\text{best}}) - E(\mathbf{x}_{r0}) + E(\mathbf{x}_{r1}) - E(\mathbf{x}_{r2})\right] = E(\mathbf{x}). \tag{4.47}$$

To calculate the variance of $\mathbf{v}$, we rewrite (4.30) as follows:

$$Var(\mathbf{v}) = (1-F_i)\mathbf{x}_{r0} + F_i(\mathbf{x}^p_{\text{best}} + \mathbf{x}_{r1} - \mathbf{x}_{r2}). \tag{4.48}$$

The two terms $\mathbf{x}_{r0}$ and $\mathbf{x}^p_{\text{best}} + \mathbf{x}_{r1} - \mathbf{x}_{r2}$ in (4.48) are independent, and they are both independent of F_i. Hence, we can apply the Lemma to calculate the variance of $\mathbf{v}$:

$$\begin{aligned}
Var(\mathbf{v}) &= E[(1-F)^2]Var(\mathbf{x}_{r0}) + E(F^2)Var(\mathbf{x}^p_{\text{best}} + \mathbf{x}_{r1} - \mathbf{x}_{r2})\\
&\quad + Var(F)\left[E(\mathbf{x}_{r0} - \mathbf{x}^p_{\text{best}} + \mathbf{x}_{r2} - \mathbf{x}_{r1})\right]^2\\
&= E[(1-F)^2]Var(\mathbf{x}) + 3E(F^2)Var(\mathbf{x})\\
&= E(1-2F+4F^2)Var(\mathbf{x}).
\end{aligned} \tag{4.49}$$

The mean and variance of $\mathbf{u}$ can be calculated directly via its definition in (4.32). However, this may lead to tedious derivations. A simpler method is to consider an auxiliary variable $\mathbf{w} = \mathbf{u} - \mathbf{x}$, i.e.,

$$w_{j,i} = \begin{cases} v_{j,i} - x_{j,i}, & \text{if } \mathrm{rand}_j(0,1) \le CR_i \\ 0, & \text{otherwise.} \end{cases} \tag{4.50}$$

Follow from basic stochastic theory, and we have

$$E(\mathbf{u}) = E(\mathbf{w}) + E(\mathbf{x}), \tag{4.51}$$

and

$$\begin{aligned}
Var(\mathbf{u}) &= Var(\mathbf{w}) + Var(\mathbf{x}) + \mathrm{Cov}(\mathbf{x},\mathbf{w}) + \mathrm{Cov}(\mathbf{w},\mathbf{x})\\
&= E(\mathbf{w}^2) - E(\mathbf{w})^2 + Var(\mathbf{x}) + \mathrm{Cov}(\mathbf{x},\mathbf{w}) + \mathrm{Cov}(\mathbf{w},\mathbf{x}).
\end{aligned} \tag{4.52}$$

According to (4.50), we have

$$E(\mathbf{w}) = \tilde{\mu}_{CR} E(\mathbf{v} - \mathbf{x}) = 0, \tag{4.53}$$

where

$$\begin{aligned} \tilde{\mu}_{CR} &= \Pr(\text{rand}_j(0,1) \le CR_i) \\ &= \textstyle\int_0^1 \Pr(\text{rand}_j(0,1) \le x | x) p_{CR_i}(x) dx \\ &= \textstyle\int_0^1 x\, p_{CR_i}(x) dx = E(CR), \end{aligned} \tag{4.54}$$

with $p_{CR_i}(x)$ denoting the pdf of CR_i. Thus, from (4.53) and (4.51), we have

$$E(\mathbf{u}) = E(\mathbf{x}). \tag{4.55}$$

The expectation of $\mathbf{w}^2$ can be easily obtained as follows:

$$E(\mathbf{w}^2) = <\boldsymbol{\Phi}, E[(\mathbf{v} - \mathbf{x})^2]>, \tag{4.56}$$

where $\boldsymbol{\Phi} = \{\phi_{j,k}\}$ is given by

$$\phi_{j,k} = \begin{cases} \Pr(\text{rand}_j(0,1) \le CR_i) = E(CR), & \text{if } j = k \\ \Pr(\text{rand}_j(0,1) \le CR_i, \text{rand}_k(0,1) \le CR_i) = E(CR^2), & \text{if } j \ne k, \end{cases} \tag{4.57}$$

with the second line obtained in a manner similar to (4.54).

Similarly, the covariance of $\mathbf{x}$ and $\mathbf{w}$ is given by

$$\begin{aligned} \text{Cov}(\mathbf{x}, \mathbf{w}) + \text{Cov}(\mathbf{w}, \mathbf{x}) &= E(\mathbf{x}\mathbf{w}') - E(\mathbf{x})E(\mathbf{w}') + E(\mathbf{w}\mathbf{x}') - E(\mathbf{w})E(\mathbf{x}') \\ &= E(\mathbf{x}\mathbf{w}') + E(\mathbf{w}\mathbf{x}') \\ &= \Pr(\text{rand}_j(0,1) \le CR_i) \{E[\mathbf{x}(\mathbf{v} - \mathbf{x})'] + E[(\mathbf{v} - \mathbf{x})\mathbf{x}]'\} \\ &= E(CR) E(\mathbf{x}\mathbf{v}' + \mathbf{v}\mathbf{x}' - 2\mathbf{x}^2), \end{aligned} \tag{4.58}$$

where the second line is obtained based on $E(\mathbf{w}) = 0$ in (4.53).

We note that $\mathbf{v}_i$ and $\mathbf{x}_i$ are independent according to (4.30) or (4.48), because all the random variables involved in the calculation of $\mathbf{v}_i$ are independent of $\mathbf{x}_i$. Furthermore, they have the same mean according to (4.47). Thus, we can rewrite the related terms in (4.56) and (4.58) as follows:

$$\begin{aligned} E[(\mathbf{v} - \mathbf{x})^2] &= E(\mathbf{v}^2) - E(\mathbf{v})E(\mathbf{x}') - E(\mathbf{x})E(\mathbf{v}') + E(\mathbf{x}^2) \\ &= E(\mathbf{v}^2) - E(\mathbf{v})E(\mathbf{v}') - E(\mathbf{x})E(\mathbf{x}') + E(\mathbf{x}^2) \\ &= \text{Var}(\mathbf{v}) + \text{Var}(\mathbf{x}), \end{aligned} \tag{4.59}$$

and

$$\begin{aligned} E(\mathbf{x}\mathbf{v}' + \mathbf{v}\mathbf{x}' - 2\mathbf{x}^2) &= E(\mathbf{x})E(\mathbf{v}') + E(\mathbf{v})E(\mathbf{x}') - 2E(\mathbf{x}^2) \\ &= 2E(\mathbf{x})E(\mathbf{x}') - 2E(\mathbf{x}^2) = -2\text{Var}(\mathbf{x}). \end{aligned} \tag{4.60}$$

Thus, (4.56) and (4.58) can be simplified as follows:

$$E(\mathbf{w}^2) = <\boldsymbol{\Phi}, \text{Var}(\mathbf{v}) + \text{Var}(\mathbf{x})>, \tag{4.61}$$

$$\mathrm{Cov}(\mathbf{x},\mathbf{w})+\mathrm{Cov}(\mathbf{w},\mathbf{x}) = -2E(CR)\mathrm{Var}(\mathbf{x}). \tag{4.62}$$

Substituting (4.53), (4.61) and (4.62) into (4.52), we have

$$\begin{aligned} Var(\mathbf{u}) &= E(\mathbf{w}^2)-E(\mathbf{w})^2+Var(\mathbf{x})+\mathrm{Cov}(\mathbf{x},\mathbf{w})+\mathrm{Cov}(\mathbf{w},\mathbf{x}) \\ &= <\Phi, \mathrm{Var}(\mathbf{v})+\mathrm{Var}(\mathbf{x})> +Var(\mathbf{x})-2E(CR)\mathrm{Var}(\mathbf{x}) \\ &= <1-2E(CR)+E(2-2F+4F^2)\Phi, Var(x)>, \end{aligned} \tag{4.63}$$

where the last line is based on the expression of Var($\mathbf{v}$) in (4.49).

Chapter 5
Surrogate Model-Based Differential Evolution

In many practical applications, optimization problems have a demanding limitation on the number of function evaluations that are expensive in terms of time, cost and/or other limited resources. Adaptive differential evolution such as JADE is capable of speeding up the evolutionary search by automatically evolving control parameters to appropriate values. However, as a population-based method by nature, adaptive differential evolution still typically requires a great number of function evaluations, which is a challenge for the increasing computational cost of today's applications. In general, computational cost increases with the size, complexity and fidelity of the problem model and the large number of function evaluations involved in the optimization process may be cost prohibitive or impractical without high performance computing resources. One promising way to significantly relieve this problem is to utilize computationally cheap surrogate models (i.e., approximate models to the original function) in the evolutionary computation.

In this chapter, JADE is extended to address optimization problems with expensive function evaluations by incorporating computationally inexpensive surrogate models and adaptively controlling the level of incorporation according to model accuracy. Radial basis function networks are used to create the surrogate models for the sake of good balance between computational complexity and model accuracy. Experimental results have shown the efficiency of adaptive incorporation of surrogate models in avoiding potential false or premature convergence while significantly reducing the number of original expensive function evaluations.

5.1 Introduction

Surrogate models have been incorporated into different evolutionary algorithms (EAs) in various ways by using history data in evolutionary computation [105], [106], [107], [108], [109], [110], [111]. They can be embedded in almost every component of evolutionary algorithms, including initialization, mutation, recombination and fitness evaluations [105]. Among these approaches, the use of surrogate models for fitness evaluations may directly and significantly reduce the number of

J. Zhang and A.C. Sanderson: Adaptive Differential Evolution, ALO 1, pp. 83–93.
springerlink.com

fitness evaluations. However, the application of surrogate models to evolutionary computation is not as straightforward as one may expect to merely obtain function values by calculating the surrogate function instead of the original function. The reason is that it is difficult to construct a surrogate model that is globally correct due to the high dimensionality, scattered distribution and limited history data. It is found that if surrogate models are used for all or most fitness evaluations, it is likely that the evolutionary algorithm prematurely converges to a false optimum which may not even be a local minimum.

Therefore, it is very essential in most cases that the surrogate model is used together with the original function evaluation. This is the issue of evolution control [105], [112] which refers to the mechanism of replacing some original function evaluations by using the surrogate model; e.g., the surrogate model may be used to predict the function values of some individuals in each generation or all individuals in some generations. In this chapter, we propose a method which conducts adaptive evolution control by estimating the accuracy of the current surrogate model in a simple and straightforward manner and then controlling the manner of utilizing the surrogate-model information according to the accuracy of the model itself.

The rest of this chapter is organized as follows. Sections 5.2 and 5.3 respectively describe the basic concepts of differential evolution and radial basis function (RBF) based surrogate model. In Section 5.4, we present the procedure of DE-AEC and explain the reasoning for the algorithm's settings. In Section 5.5, we compare DE-AEC with other DE algorithms without adaptive evolution control. Finally, we summarize concluding remarks in Section 5.6.

5.2 Adaptive Differential Evolution

The incorporation of the surrogate model with DE, especially adaptive DE algorithms, raises an interesting research topic in that: (i) adaptive DE retains the simplicity of classic DE but shows better convergence performance for a large set of test problems compared to classic DE and other EAs investigated [21], [22], and (ii) contrary to the case of classic DE, the control parameters of adaptive DE are usually problem-insensitive and thus avoid lengthy trial and error which is obviously impractical in the case of expensive function evaluations. Indeed, it is very meaningful to investigate the performance benefit of adaptive evolution control over an already improved adaptive DE algorithm, such as jDE in [22] and JADE proposed in the Chapter 4.

In this chapter, we consider JADE as the basis of our adaptive evolution control based algorithm, DE-AEC. However, as it becomes clear below, the scheme of adaptive evolution control can also be easily incorporated into other classic or adaptive DE algorithms. For this approach, JADE (instead of classic DE such as DE/rand/1/bin,) is chosen as the underlying scheme of adaptive evolution control based on the following considerations.

First, the control parameters, F and CR, in classic DE are usually problem-dependent and thus have to be tuned by a trial and error method, which inevitably requires a large number of optimization trials. This is not practical in the case of expensive function evaluations. If parameters are not well tuned due to the limit of trials, classic DE may lead to frequent premature or false convergence which obscures the benefit of the surrogate model. On the contrary, adaptive DE algorithms such as JADE benefit from their problem-insensitive properties by adaptively updating control parameters during the evolutionary process. The benefit of a surrogate model can thus be clearly observed with less effect of false convergence. Furthermore, adaptive algorithms usually have much better convergence performance in terms of both rate and reliability. It is thus very interesting to investigate the benefit of DE-AEC over an already improved DE algorithm for optimization problems with expensive function evaluations.

For notational simplicity, we omit the superscript g or $g+1$ in the rest of the chapter if no confusion is possible.

5.3 RBF Surrogate Model

We now provide a brief description of the RBF surrogate model (with cubic basis function) that was extensively studied by Powell [113], [114]. Assume that we are given k distinct points $\mathbf{x}_1, \mathbf{x}_2, \ldots, \mathbf{x}_k \in \mathrm{R}^D$, where the true function values, $f(\mathbf{x}_i)$, are known. In this method, we use an interpolant of the form

$$s(\mathbf{x}) = \sum_{i=1}^{k} w_i \phi(|\mathbf{x} - \mathbf{x}_i|) + \mathbf{b}^T\mathbf{x} + a, \mathbf{x} \in \mathrm{R}^D, \tag{5.1}$$

where $\mathbf{w} = (w_1, w_2, \ldots, w_k)^T \in \mathrm{R}^k$, $\mathbf{b} \in \mathrm{R}^D$, $a \in \mathrm{R}$, and ϕ is a cubic basis function, i.e., $\phi(r) = r^3$ for $r \geq 0$.

The unknown parameters $\mathbf{w}$, $\mathbf{b}$, and a are obtained as the solution of the system of linear equations

$$\begin{pmatrix} \Phi & \mathrm{P} \\ \mathrm{P}^T & 0 \end{pmatrix} \begin{pmatrix} \mathbf{w} \\ c \end{pmatrix} = \begin{pmatrix} F \\ 0 \end{pmatrix}, \tag{5.2}$$

where Φ is the $k \times k$ matrix with $\Phi_{i,j} = \phi(\|x_i - x_j\|_2)$, $\mathbf{c} = (\mathbf{b}^T, \mathrm{a})^T$, $\mathbf{F} = (f(\mathbf{x}_1), f(\mathbf{x}_2), \ldots, f(\mathbf{x}_k))^T$, and

$$\mathrm{P}^T = \begin{pmatrix} \mathbf{x}_1 & \mathbf{x}_2 & \cdots & \mathbf{x}_k \\ 1 & 1 & \cdots & 1 \end{pmatrix}. \tag{5.3}$$

In [113], [114], Powell shows that if the rank of P is $D + 1$, then the square matrix in the left hand of (5.2) is nonsingular. Thus, the linear system (5.2) has a unique solution and we can use (5.1) as the approximate model to predict function values.

In our research, the RBF surrogate model is constructed for each newly generated offspring individual using only the $k = D^2$ nearest data points in the database of the $k' = 10D^2$ best-so-far evaluated points. This is consistent with the fact that the neighboring points are likely to have more impact than remote ones [115]. Note that k is roughly proportional to the minimum number, $(D+1)(D+2)/2$, of data points required to fit a quadratic model, while k' is set to be moderate considering the possible limit on database size and the complexity of searching for the nearest data points of each individual.

5.4 DE-AEC: Differential Evolution with Adaptive Evolution Control

In this section, we elaborate the basic procedure of DE-AEC and explain the reasoning of the algorithm's settings.

5.4.1 *Procedure of DE-AEC*

As shown in Figure 5.1, DE-AEC follows the basic procedure of usual differential evolution algorithms described in Chapter 2.2, except that each parent individual $\mathbf{x}_i$ generates K offspring ($\mathbf{u}_{i,j}$, j = 1, 2, ..., K) and then chooses the promising one as $\mathbf{u}_i$ to compete with itself in the latter selection. To be specific, each of the K individuals is independently generated by the standard mutation and crossover. The $\mathbf{u}_i$ is then selected from them in a manner explained below. It is compared with $\mathbf{x}_i$ based on their true function values $f(\mathbf{u}_i)$ and $f(\mathbf{x}_i)$. That is, we conduct NP true function evaluations at each generation, as in other DE algorithms.

We next explain the distinct feature of DE-AEC, i.e., its procedure to choose the competitive offspring $\mathbf{u}_i$ based on the function values predicted by the surrogate model and the accuracy of the model as well. The accuracy of the surrogate model is defined in a way consistent with the selection procedure of DE, where the ranking of the two function values matters. Denote r_{si} and r_{fi} as the rankings of $s(\mathbf{u}_i)$ and $f(\mathbf{u}_i)$ in their respective vectors $\mathbf{s} = (s(\mathbf{u}_1), s(\mathbf{u}_2), \ldots, s(\mathbf{u}_{NP}))$ and $\mathbf{f} = (f(\mathbf{u}_1), f(\mathbf{u}_2), \ldots, f(\mathbf{u}_{NP}))$. The accuracy metric refers to the correlation coefficient ρ of the two rank vectors $\mathbf{r}_s = (r_{s1}, r_{s2}, \ldots, r_{s,NP})$ and $\mathbf{r}_f = (r_{f1}, r_{f2}, \ldots, r_{f,NP})$[1]. For simplicity, we regard the accuracy levels as *poor, moderate* or *excellent* when $\rho \le \rho_1^*$, $\rho_1^* < \rho < \rho_2^*$ or $\rho \ge \rho_2^*$ for certain constants ρ_1^* and ρ_2^* $(-1 < \rho_1^* < \rho_2^* < 1)$, respectively. An appropriate setting is that ρ_1^* =1/3 and ρ_2^* = 2/3.[2]

Among the K offspring $\mathbf{u}_{i,j}$ of $\mathbf{x}_i$, denote the one having the best approximate function value as $\mathbf{u}_{i*}$, i.e., $s(\mathbf{u}_{i*}) = \min_j s(\mathbf{u}_{i,j})$. Then, we set either $\mathbf{u}_i = \mathbf{u}_{i*}$ (*elitist choice*) or $\mathbf{u}_i = \mathbf{u}_{i,1}$ (*random choice*) in the following manner.

[1] Other common accuracy metrics include root mean square error and correlation coefficient between $\mathbf{s}$ and $\mathbf{f}$ [107]. These metrics, however, suffer from the usual situation where function values spread over several magnitudes so that a few of them dominate the calculation of the metric.

[2] In practice, the RBF model used in this chapter rarely leads to a large negative accuracy ρ (i.e., close to -1), even in the case of the most difficult problem investigated.

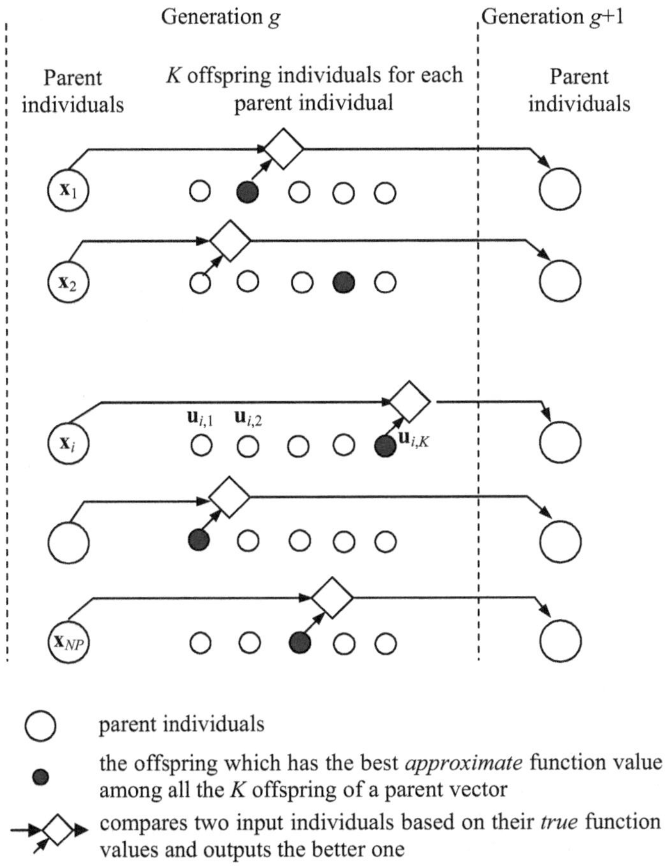

Fig. 5.1 An illustration of the procedure of DE-AEC (K = 5). In DE-AEC, the trial vector $\mathbf{u}_i$ is chooses as either the first randomly generated offspring individual $\mathbf{u}_{i,1}$ or the one $\mathbf{u}_{i*}$ which has the best *approximate* fitness value among the K offspring. Note that $\mathbf{u}_{i*}$ could be the same as $\mathbf{u}_{i,1}$. The trial vector is then compared with the corresponding parent vector based on their *true* fitness values. The better one survives and is as a parent vector in the next generation

Denote M as the number of $\mathbf{u}_i$'s that are set to be the respective $\mathbf{u}_{i*}$, and$S$ as the set of indices of these $\mathbf{u}_i$'s. The number M is given by $M = [q \cdot NP]$, where $[\cdot]$ is the round operation, and q is (roughly) the percentage of elitist choices. q is initialized to be 1 and then updated at each generation as follows:

$$q = \begin{cases} \max\{q^*, q - \beta(\rho - \rho_1^*)\}, & \text{if } \rho \geq \rho_2^* \\ \max\{q^*, q + \beta(\rho - \rho_1^*)\}, & \text{if } \rho < \rho_2^*, \end{cases} \quad (5.4)$$

where β and q^* are positive constants. Given its size M, the set S is then calculated

$$S = \begin{cases} \{i|s(y_i) \leq s_M\}, & \text{if } \rho \geq \rho_2^* \\ \{i|\pi_i \leq M\}, & \text{if } \rho < \rho_2^*, \end{cases} \tag{5.5}$$

where π_i is the position of i in a random permutation of the set $\{1, 2, \ldots, NP\}$ and s_M is the M-th smallest element of $\mathbf{s}$. Finally, the trial individual $\mathbf{u}_i$ is generated as follows:

$$\mathbf{u}_i = \begin{cases} \mathbf{u}_{i*}, & \text{if } i \in S, \\ \mathbf{u}_{i,1}, & \text{if } i \notin S. \end{cases} \tag{5.6}$$

In both (5.4) and (5.5), different operations are conducted depending on whether the accuracy of the surrogate model is excellent or not. In (5.4), we respectively decrease or increase the percentage q, while always keeping it greater than a lower limit q^*. In (5.6), we choose the indices of $\mathbf{u}_i = \mathbf{u}_{i*}$ according to approximate function values or merely in a random manner.

5.4.2 Explanation of q, S and K

The values of p and S are updated differently depending on whether the approximation of surrogate model is excellent ($\rho \geq \rho_2^*$). The difference comes from an essential observation: The approximation becomes excellent when the population congregates closely or approaches the basin around a global or local minimum. Special measures should be taken to avoid false convergence to a local minimum or other points.

First, consider the case of excellent approximation, i.e., $\rho \geq \rho_2^*$. On the one hand, to avoid the aforementioned false convergence, we encourage more random choice of $\mathbf{u}_i = \mathbf{u}_{i,1}$ by decreasing q to $q - \beta(\rho - \rho_1^*)$. On the other hand, a lower limit of q, $q \geq q^*$, is necessary to take advantage of the high quality of approximation in order to speed the evolutionary progress. In practice, a reduced q does not affect the progress rate much as long as $q \geq q^*$ for an appropriate q^*, but it can significantly improve the probability of correct convergence. More interestingly, it is observed that a smaller q even improves the progress rate in some cases (e.g., the Rosenbrock function f_5) if elitist choices tend to concentrate all individuals to a small niche, causing a trivial progress rate. The first line of (5.5) aims to choose the M most promising offspring based on their approximate function values. This is beneficial when the approximation is reliable.

Second, consider the case of poor or moderate approximation when the premature convergence is not a serious problem. In this case, the update of q follows a simple principle: the higher (lower) the accuracy level, the more (fewer) elitist choices we have. This corresponds to the operation $q = q + \beta(\rho - \rho_1^*)$. In addition, we set the lower limit q^* for q because the approximate model usually provides more helpful information than misleading information even in the case of poor approximation. Furthermore, we randomly choose the M positions of elitist choice, because the surrogate model is not very reliable. This is the operation in the second line of (5.5).

In regard to K, it balances the effectiveness and the reliability of using the surrogate model. The benefit of the surrogate model is trivial if it is close to 1. If it is very large, on the other hand, each individual approaches its adjacent local minimum faster and may lead to frequent premature convergence.

5.5 Performance Evaluation

In this section, DE-AEC is applied to minimize a set of fifteen benchmark functions including all Dixon–Szegö test functions $f_{14} - f_{20}$ in Table 4.4 and some function in Table 4.3: spherical function f_1, ellipsoid function f_3, Rosenbrock function f_5, Rastrigin function f_9, Ackley function f_{10}, Griewank function f_{11}, and penalized functions f_{12} and f_{13}.

DE-AEC is compared with JADE as well as DE/rand/1/bin. For fair comparison, we follow the parameter settings of JADE in Chapter 4 (i.e., $c = 0.1$ and $p = 0.05$) and the parameters of DE/rand/1/bin are set to be $F = 0.5$ and $CR = 0.9$, as used in [22] and recommended by other references [2], [9], [10]. For all DE algorithms, we set the population size NP to be $10D$ for $D < 10$ and $3D$ for $D = 10$. In addition, different from the usual initialization procedure of JADE and DE/rand/1/bin (as described in Section II), DE-AEC generates its initial population by choosing the best NP vectors from a SLHD (symmetric Latin hypercube design [116]) of size $15D$ based on true function evaluations. In all simulations, the parameters of DE-AEC are set to be $K = 5$, $\beta = 1/4$, $q^* = 1/3$, $\rho_1^* = 1/3$ and $\rho_2^* = 2/3$.

5.5.1 *Success Rate and Success Performance*

Simulation results are obtained for each algorithm based on 50 independent runs. Table 5.1 summarizes the success rate (SR) and success performance (SP) [117] of each algorithm. SR refers to the rate of runs where an accuracy of ε is achieved (i.e., the best-so-far function value is at most ε from the global minimum value) before 100,000 true function evaluations. Note that this number is set to be sufficiently large for the purpose of distinguishing slow convergence from premature convergence. In [117], SP is evaluated as a quantity: the number of true function evaluations (to achieve an accuracy ε) averaged over successful runs $\times$ the number of total runs / the number of successful runs. As shown in Table 5.1, the success performance is calculated for both low-level ($\varepsilon_1 = \varepsilon_2 = 10^{-2}$) and high-level ($\varepsilon_1 = \varepsilon_2 = 10^{-8}$) accuracy requirements.

In addition, the median of the difference between the best-so-far function values and the optimum value is plotted in Figure 5.2 – Figure 5.4 for an illustrative comparison.

5.5.2 *Simulation Results*

Before comparing DE-AEC with other algorithms, we first make a preliminary observation on DE/rand/1/bin and JADE. It is clear from Table 5.1 and

Table 5.1 A comparison of success rate (SR) and success performance (SP) for different accuracy requirements

Functions			f_1	f_3	f_5	f_9	f_1	f_{11}	f_{12}	f_{13}	f_{14}	f_{15}	f_{16}	f_{17}	f_{18}	f_{19}	f_{20}
DE-AEC	ε_1	SR	100	100	94	100	100	96	100	100	100	100	100	88	94	96	100
		SP	**1494**	**3261**	**8484**	**5825**	**1914**	**6439**	**1668**	**1676**	**264**	**211**	**185**	**1263**	**1114**	**990**	**987**
	ε_2	SR	100	100	94	100	100	96	100	100	100	100	100	88	94	96	100
		SP	**3089**	**6967**	**1.1E4**	**8737**	**5135**	**1.3E4**	**3823**	**3958**	**786**	**512**	**866**	**4836**	**2436**	**2220**	**2094**
JADE	ε_1	SR	100	100	86	98	100	98	100	100	100	100	100	92	86	98	98
		SP	2675	5043	1.7E4	8515	3412	1.3E4	2315	2444	420	439	394	2838	3169	2372	2581
	ε_2	SR	100	100	86	98	100	98	100	100	100	100	100	76	84	98	96
		SP	5566	9082	2.1E4	1.3E4	9451	2.5E4	5362	5620	1133	999	1503	8940	6009	4658	4839
DE/rand/1/bin	ε_1	SR	100	76	0	8	100	20	100	98	100	100	100	46	96	96	100
		SP	4004	1.1E4	–	4.0E5	4959	5.7E4	3260	3835	408	408	398	7112	2464	2219	2190
	ε_2	SR	98	32	0	8	100	6	100	98	100	100	100	46	96	96	100
		SP	8100	7.1E4	–	4.6E5	1.3E4	2.1E5	7051	7697	998	981	1616	1.8E4	4668	4457	4318

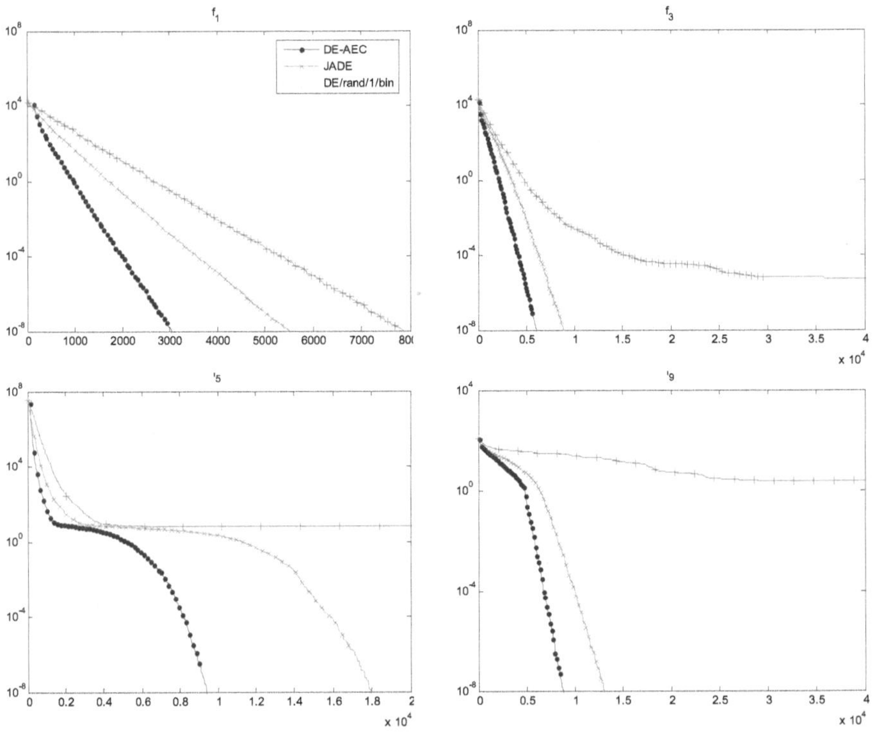

Fig. 5.2 Convergence graph (the median of 50 independent runs) for f_1, f_3, f_5, and f_9. The horizontal axis is the number of true function evaluations. The vertical axis is the difference between the best-so-far function value and the optimal value

Figure 5.2 – Figure 5.4 that DE/rand/1/bin is not satisfactory in terms of success rate , especially for high dimensional problems. Without tedious (and expensive) parameter tuning, the control parameters adopted by DE/rand/1/bin are not well suited for many of these problems. JADE, on the other hand, shows more reliable performance, and its success performance is comparable to DE/rand/1/bin for low-dimensional problems and much better for high-dimensional problems. It is thus suitable to serve as the basis of DE-AEC that aims at optimization problems with expensive function evaluations.

We next compare the results obtained by JADE and DE-AEC. As summarized in Table 5.1, the success performance of DE-AEC is significantly better than JADE for either low-level ($\varepsilon_1 = 10^{-2}$) or high-level ($\varepsilon_2 = 10^{-8}$) accuracy requirement for fourteen out of the fifteen test functions. For this set of functions, the success performance implies that a 25 – 65 % (compared to JADE) or 25 – 90 % (compared to DE/rand/1/bin) reduction of the number of true function evaluations is achieved on average.

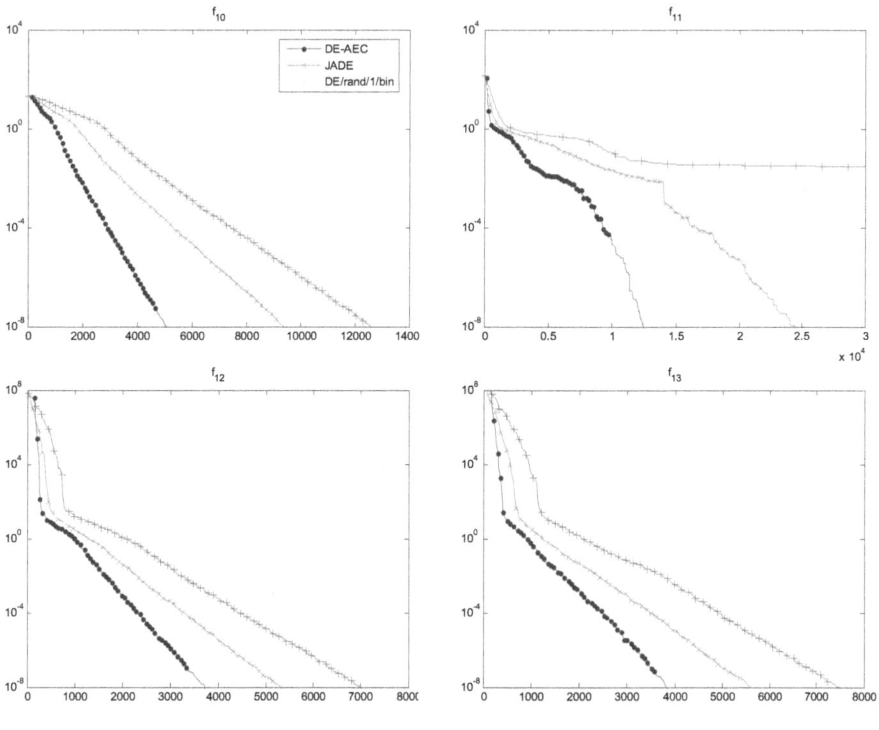

Fig. 5.3 Convergence graph (the median of 50 independent runs) for $f_{10} - f_{13}$. The horizontal axis is the number of true function evaluations. The vertical axis is the difference between the best-so-far function value and the optimal value

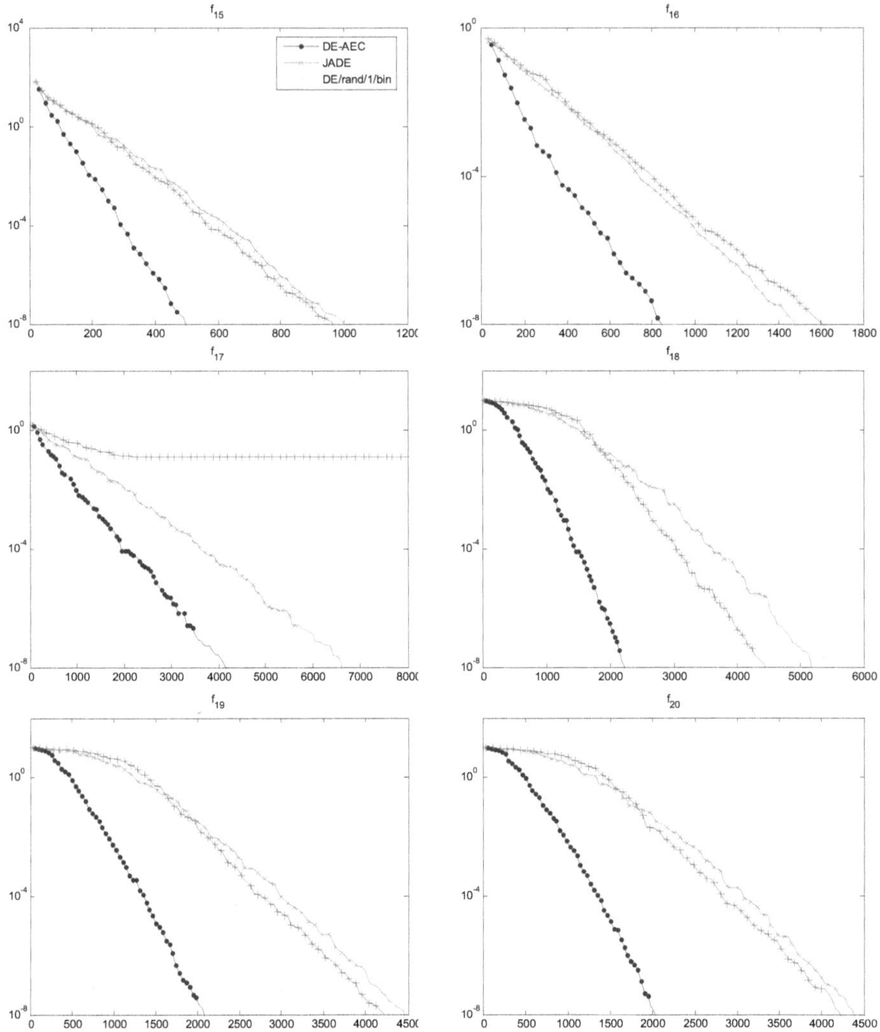

Fig. 5.4 Convergence graph (the median of 50 independent runs) for $f_{15} - f_{20}$. The horizontal axis is the number of true function evaluations. The vertical axis is the difference between the best-so-far function value and the optimal value

5.6 Summary

We have proposed an adaptive evolution control based differential evolution algorithm, DE-AEC, to significantly reduce the computation cost of expensive function optimization. Taking the adaptive differential evolution algorithm proposed in Chapter 4 as the basis, DE-AEC generates multiple offspring for each parent vector, among which one is selected to compete with the parent based on the information (the predicted function value and the accuracy) of the surrogate model. The accuracy of the surrogate model is also used as an indicator of potential false convergence and special operations are performed to improve the convergence reliability of the algorithm. As verified by simulation, DE-AEC reduces the number of true function evaluations by 25% – 90% in the achievement of either low-level (10^{-2}) or high-level (10^{-8}) accuracy.

It is worth noting that our current studies are meaningful only when the computational cost of the surrogate-model is much less than that of the true function evaluation. While this is the situation in many practical applications, careful considerations should be taken if the assumption does not hold. In that case, it will be a meaningful research topic to study the tradeoff between the accuracy and the computational complexity of different surrogate models and to compare different schemes of integrating the surrogate model into evolutionary computation.

Chapter 6
Adaptive Multi-objective Differential Evolution

Multi-objective optimization exists everywhere in real-world applications such as engineering, financial, and scientific applications, because the outcome is directly linked to cost, profit and/or many other criteria that have heavy impacts on performance, safety, environment, etc. It is difficult to provide an ultimate comparison among different outcomes via only one dimension, as the involved multiple criteria/objectives are generally competing and non-commensurable. For example, a financial manager needs to take both return and risk into consideration when making an investment decision; an air traffic controller needs to consider both the reduction of system-level airspace congestion and the satisfaction of different stakeholders' preferences.

In this chapter, we investigate the application of parameter adaptive differential evolution to multi-objective optimization problems. To address the challenges in multi-objective optimization, we create an archive to store recently explored inferior solutions whose difference with the current population is utilized as direction information about the optimum, and consider a fairness measure in calculating crowding distances to prefer the solutions whose distances to the nearest neighbors are large and close to be uniform. As a result, the obtained solutions can spread well over the computed non-dominated front and the front can be moved fast toward the Pareto-optimal front. The control parameters of the algorithm are adjusted in an adaptive manner, avoiding parameter tuning for problems of different characteristics.

6.1 Introduction

Many real-world applications involve multi-objective optimization, because they are linked to competing or conflicting outcomes such as profit, risk and/or many other performance related criteria. The presence of multiple objectives introduces extra challenges to researchers and more efforts are required for these problems beyond traditional techniques like linear and nonlinear programming as well as single-objective evolutionary algorithms. According to the interaction with a decision-making process, the multi-objective optimization can be classified as follows [53], [54]:

J. Zhang and A.C. Sanderson: Adaptive Differential Evolution, ALO 1, pp. 95–113.
springerlink.com

1. *Prior preference articulation*: The multiple objectives are linearly or nonlinearly aggregated into a scalar function based on prior preference structure from decision makers. This transforms the problem into a single-objective problem prior to optimization.
2. *Progressive preference articulation*: The optimization is intertwined with decision making process when partial preference information is available.
3. *Posterior preference articulation*: A family of tradeoff solutions is found before a decision is made to determine the best solution.

The aggregation method in prior preference articulation is mainly applied in the early development of multi-objective evolutionary algorithms [55]. While the progressive preference articulation is also a current research interest [56], the posterior preference has been gaining significant attention from researchers in proposing multi-objective evolutionary algorithms (MOEAs) to search for tradeoff solutions [14], [15], [34], [35], [57] – [61], [119] – [134].

There are two key issues in the design of multi-objective evolutionary algorithms: to maximize the diversity and spread of solutions, and to increase the speed of convergence to the Pareto-optimal front. In the literature, the first issue is addressed by various diversity maintenance mechanisms that are mainly based on the estimation of crowding density of solutions. For example, two commonly used crowding estimation techniques are proposed in SPEA2 [60] and NSGA-II [35]. In SPEA2, a decreasing function of the distance to the k-th nearest neighbor is used to estimate the crowding density of a solution. In NSGA-II, a crowding distance is calculated as the summation of distances of two points on either side of a solution along each dimension in the objective space. In both methods, solutions with smaller crowding distances or higher crowding densities are considered inferior, and thus are more likely pruned from the population to improve the diversity.

One method to address the second issue is parameter tuning or parameter control. The former requires tedious trial and error, while the latter is "a very interesting topic that very few researchers have addressed in the specialized literature" of MOEAs [119]. A second method is to integrate the promising features in existing MOEAs into recently developed single objective evolutionary algorithms (SOEAs) that have shown great success. However, the reason for the success of SOEAs may become invalid in the case of multi-objective optimization. For example, in SOEAs, it has been shown that the best solutions carry direction information towards the optimum and thus can be utilized to speed up convergence. This advantage becomes invalid or less significant in multi-objective optimization, because all or most solutions in the population may become non-dominated from one another after only a few generations, especially when the dimension of the objective space is very high. In this case, as illustrated later in Figure 6.1, the best solutions are identified according to their secondary metric of crowding density, instead of the Pareto dominance rank, and thus do not necessarily indicate the progress direction to the Pareto-optimal front.

Addressing the above issues, we propose in this chapter an adaptive multi-objective differential evolution algorithm. Its control parameters are adjusted in an adaptive manner, and thus avoid users' interaction before and during the optimization process and avoid degraded performance due to inappropriate parameter settings. An external archive is introduced to store recently explored inferior solutions whose difference with the current population indicates a promising direction towards the Pareto-optimal front. It is worth noting that this is different from the archives used in many existing MOEAs to store the best-so-far non-dominated solutions. In addition, we calculate the crowding distance according to a fairness measure to prefer the solutions whose distances to the nearest neighbors are large and close to be uniform. This is helpful to better spread solutions over the computed non-dominated front.

The rest of this chapter is organized as follows. Section 6.2 briefly reviews several classic multi-objective evolutionary algorithms and introduces existing studies of differential evolution in multi-objective optimization. Section 6.3 analyzes the challenges in multi-objective optimization and proposes new methods of estimating crowding density and extracting direction information from archived inferior solutions. These features, together with the adaptive parameter control in JADE, compose the basic procedure for multi-objective optimization. Simulation results are presented in Section 6.4 to demonstrate the performance of JADE. Finally, concluding remarks are presented in Section 6.5.

6.2 Multi-objective Evolutionary Algorithms

This section briefly reviews several classic Pareto-dominance based multi-objective evolutionary algorithms and the applications of differential evolution in this area. It provides a convenient way to compare the diversity maintenance mechanisms and archive techniques in these algorithms with those proposed in JADE. Interested readers are referred to [55] [57] for comprehensive reviews of MOEAs.

6.2.1 PAES

Knowles and Corne [58] proposed the Pareto archived evolution strategy (PAES) with a reference archive of non-dominated solutions found in the evolution. The simplest form of PAES employs $(1+1)$ evolution strategy and uses a mutation operator for local search. The archive approximates the Pareto front and identifies the fitness of current and potential solutions. In addition, a map of grid is applied in the algorithm to maintain the diversity of the archive.

6.2.2 SPEA2

The SPEA2 [60] is an improved version of SPEA, the strength Pareto evolutionary algorithm (SPEA) [58]. Both the archive and population in the algorithm are assigned a fitness based on the strength and density estimation. The strength of an individual is denoted as the number of individuals that dominates it, while the density estimation is based on the k-th nearest neighbor method [118] where the density of any solution is a decreasing function of the distance to the k-th nearest neighbor. A truncation method based on the density is applied to keep the archive at a fixed size.

6.2.3 PESA

The Pareto envelope-based selection algorithm (PESA) [61] is motivated from SPEA and PAES. It uses an external population to store the computed Pareto front and an internal population to generate new candidate solutions. The PESA maintains a hyper grid based scheme to keep track of the degree of crowding in different regions of the archive, which is applied to maintain the diversity of external population and to select the internal population from the external population.

6.2.4 NSGA and NSGA-II

Srinivas and Deb [120] developed an approach called non-dominated sorting genetic algorithm (NSGA). The fitness assignment is carried out in several steps. In each step, the solutions constituting a non-dominated front are assigned the same dummy fitness value. Then, these solutions are ignored in the following step and the next non-dominated front is extracted. This procedure repeats until all individuals in the population are classified. In [120], this fitness assignment method was combined with a stochastic remainder selection.

In [35], Deb et al. proposed NSGA-II as an improved version of NSGA. It uses elitism and a comparison operator that ranks the population based on Pareto dominance ranking and crowding density at a lexicographic order: In the case of a tie in dominance rank, the individual with a lower density succeeds in the selection. The dominance ranking has the same meaning as in NSGA but is calculated in a more efficient manner. The crowding distance of a solution is calculated as the sum of distances to two adjacent points on its either side along each dimension in the objective space. The comparison operator makes the NSGA-II considerably faster than NSGA while producing very good results.

In [35], NSGA-II is shown to outperform PESA and SPEA in terms of finding a diverse set of solutions and in converging near the true Pareto front. In [121], Khare et al. compared the scalability of NSGA-II, PESA, and SPEA2, by considering different numbers of objectives (2 to 8). The results show that PESA is the best in terms of converging to the Pareto-optimal front but it has poor diversity maintenance. SPEA2 and NSGA-II perform equally well in terms of convergence and diversity maintenance whereas NSGA-II runs much faster.

6.2.5 Differential Evolution Based MOEAs

As a relatively recent development in the field of evolutionary algorithms, differential evolution has shown its success in various single objective optimization problems. The extensions of DE to multi-objective optimization, such as PDE [122], [123], [124], Madavan's approach [125] , MODE [126], [15] , DEMORS [127], and NSDE [128], DEMO [129], [130], and GDE3 [34], incorporate successful features in existing MOEAs and also introduce various new operations suitable to multi-objective optimization.

PDE [122], [123], [124] was the first approach of applying DE to multi-objective optimization. It employs DE to create new individuals and then keeps only the non-dominated ones as the basis for the next generation. PDE outperforms SPEA on two test problems. Like PDE, Madavan [125] applies DE to create offspring individuals and then combines parents and offspring to calculate the non-dominated rank and diversity rank (in terms of the crowding distance metric) for each individual. Two variants were investigated in [125]. In the first one, a one-to-one comparison between each pair of offspring and parent vectors is conducted and the child succeeds if and only if it has a higher non-dominated rank or a higher diversity rank (if their ranks are the same). In the second one, all offspring and parent vectors are put together and the best ones are selected into the next generation according to the non-dominated rank and diversity rank (similar to the method of NSGA-II). The latter variant has proven to be very efficient and produced favorable results for several problems. The algorithm DEMO [129] [130] is similar to Madavan's approach but is implemented in three variants (DEMO/parent, DEMO/closest/dec and DEMO/closest/obj). Compared to NSGA-II and PDEA, DEMO achieves competitive results on five ZDT test problems. Xue et al. [126], [15] introduced the Multiobjective Differential Evolution (MODE) algorithm. This algorithm also uses the Pareto-based ranking assignment and the crowding distance metric, but in a different manner than Madavan's approach. In MODE, the fitness of an individual is first calculated using Pareto-based ranking and then reduced with respect to the individual's crowding distance value. This single fitness value is then used to select the best individuals for the new population. MODE was tested on five benchmark problems where it produced better results than SPEA. In [133], GDE3, when using the product distance as a crowding estimation metric and pruning the most crowded solutions one-by-one, obtained solutions that spread clearly better than NSGA-II and SPEA2 while other performance metrics are competitive.

In [131], Zielinski and Laur compared a set of DE variants for multi-objective optimization, where (i) we *may* or *may not* conduct a one-to-one comparison between each pair of parent and offspring vectors (i.e., the original DE selection scheme) before the dominance ranking and crowding density sorting among all solutions, and (ii) the most crowded solutions are pruned from the population *all at once* or *one by one*. It is shown [131], [34] that the original DE selection scheme is beneficial for performance improvement and enables a multi-objective algorithm to fall back to the original DE algorithm if there is a single objective. Also, it is helpful to improve the population diversity by removing the most crowded solutions one by one and update the crowding density estimation after each removal [132]. In addition, different methods have been proposed to calculate the crowding density in the case of three or more objectives. It is shown that, compared to previous methods (e.g., the one in NSGA-II), the product [133] or harmonic mean [134] of the distances to the nearest neighbors may serve as an effective, though relatively complex, crowding density estimation metric to improve population diversity.

6.3 JADE for Multi-objective Optimization

In differential evolution, the crossover can be conducted in the same way for single- and multi-objective optimization. However, the selection and mutation need to be reconsidered because of the challenges in multi-objective optimization. First, the selection operation requires the comparability between any distinct solutions in multi-dimensional objective space. This, however, cannot be guaranteed by a Pareto dominance comparison as it defines a partial order among solutions. A secondary metric such as the crowding density needs to be introduced to compare two solutions if the dominance comparison ties.

Second, the best-solution information used in a greedy mutation strategy loses its original meaning of providing direction information to the optimum. This situation usually appears after only a few generations, especially when the dimension of the objective space is high and thus most or all solutions in the population quickly become non-dominated from one another. In this case, the best solutions are identified mainly according to their crowding density, instead of the dominance ranking. As a result, a usual greedy mutation strategy is useful to move the population towards the sparsest regions which however are not necessarily closer to the true Pareto front.

6.3.1 Pareto Dominance and Crowding Density

According to the definition in Chapter 2.4, a solution vector $\mathbf{f} = (f_1, f_2, \ldots, f_M)$ is said to dominate another vector $\mathbf{g} = (g_1, g_2, \ldots, g_M)$ in the objective space, if and only if all components of **f** are no worse than the components of **g** and at least one component of **f** is better than that of **g**, i.e.,

$$\forall i \in \{1,2,...,M\} : f_i \le g_i \text{ and } \exists i \in \{1,2,...,M\} : f_i < g_i. \tag{6.1}$$

If it dominates, the solution vector **f** (and its corresponding vector in the decision variable space) is considered better than the other vector **g** (and its corresponding decision vector). It is clear that two objective vectors may be incomparable (i.e., not dominated by each other) according to the definition of Pareto dominance.

A solution is considered as Pareto optimal if it is not dominated by any other vectors in the objective space. The set of these non-dominated solutions composes the Pareto-optimal front. In general, there are a huge, if not infinite, number of Pareto optimal solutions, while an evolutionary algorithm can only search for a small set of representative ones that are expected to be well diversified and to spread over the Pareto-optimal front. In the literature, this is usually achieved by different techniques of estimating the crowding density of solutions. Solutions that are more crowded are more likely to be pruned from the population to maintain the diversity.

6.3.2 Selection

In the selection operation of JADE, solutions are compared according to the Pareto dominance and crowding density estimation at a lexicographic order. That is, a solution is considered better than another one if it dominates or if it has a lower crowding density when the dominance comparison ties. To be specific, JADE adopts a three-step comparison to select NP vectors from the pool $\mathbf{P}'$ of $2NP$ parent and trial vectors. The selected ones become the parent population in the next generation.

In the first step, a one-to-one dominance comparison is conducted between each pair of parent and trial vectors. If one dominates, the other one is immediately removed from $\mathbf{P}'$ This is similar to the original one-to-one comparison of DE in single-objective optimization. After this step, the size of $\mathbf{P}'$ ranges from NP to $2NP$.

The second step is to further reduce the size of $\mathbf{P}'$ based on the dominance relationship among all solutions. First, all non-dominated solutions in $\mathbf{P}'$ are assigned rank 1, indicating the best fitness in the population. Then, the remaining solutions are considered and the non-dominated ones among them are assigned rank 2. This process repeats until no less than NP vectors have been assigned rank values. All other vectors are immediately removed from $\mathbf{P}'$.

The third step is based on the crowding density estimation on all solutions that have the worst rank value. The most crowded among them is pruned one by one, with the crowding density of remaining solutions updated after each removal until the size of $\mathbf{P}'$ is reduced to NP. The density estimation is based on the consideration that the crowdedness of a solution is lower if its distances to nearest neighbors are *larger* and *closer to be uniform*. For this purpose, the fairness measure that is usually used in network engineering [135] to maintain the fairness (uniformity) of competing variables (e.g., network flow rates) while maximizing their summation can serve as a good method of estimating the crowding density. Consider the (p, α)-proportionally fairness measure in [135]. With a usual $p = 1$ and $\alpha \geq 0$, we can calculate the crowding distance d_i of a solution i as

$$d_i = \begin{cases} \sum_{i=1}^{k} \log d_{ij}, & \text{if } \alpha = 1 \\ (1-\alpha)^{-1} \sum_{i=1}^{k} d_{ij}^{1-\alpha}, & \text{otherwise,} \end{cases} \tag{6.2}$$

where d_{ij} is the Euclidean distance of solution i to its j-th nearest neighbor in $\mathbf{P}'$. It is easy to show that if $\alpha = 0$, d_i is the sum distance. If $\alpha = 1$, it is *equivalent* to the product distance,

$$d_i = \prod_{i=1}^{k} d_{ij}; \tag{6.3}$$

if $\alpha = 2$, the harmonic distance

$$d_i = \frac{1}{\frac{1}{d_{i,1}} + \frac{1}{d_{i,2}} + \cdots + \frac{1}{d_{i,2(M-1)}}}; \tag{6.4}$$

and if $\alpha \to \infty$, the minimum distance

$$d_i = \min(d_{i,1}, d_{i,2}, ..., d_{i,k}). \tag{6.5}$$

It is worth noting that the harmonic distance and product distance have been respectively adopted in MODE [134] and GDE3 [133] as the crowding distance, while the minimum distance is a special case of the k-nearest neighbor method used in SPEA2 [60]. Experimental results show that the sum distance and minimum distance usually lead to less satisfactory results, while both the product distance and harmonic distance perform well for the test problems studied in this chapter.

As GDE3 is considered as a benchmark algorithm in this chapter, we adopt the product distance (6.3) in JADE for fair comparison. Different from the setting in GDE3 (where $k = M$), however, we consider $k = 2(M-1)$ nearest neighbors in calculating the crowding distance. This number is selected based on the consideration that the dimension of the Pareto front is usually $M - 1$ and each solution is expected to have two nearest neighbors on its either side along each dimension.

6.3.3 Mutation

In multi-objective optimization, most or all individuals in the population usually become non-dominated from one another after only a few generations. In this case, the best solutions are mainly identified according to the crowding density. Thus, in a usual greedy strategy, the population is moved towards the sparsest region where the best solutions are located. This is different from the situation of single-objective optimization or in the early stage of multi-objective optimization where the best solutions indicate the direction towards the optimal solutions. A simple illustration is shown in Figure 6.1. It is clear that the best solutions are incapable of indicating promising directions when they become non-dominated from many other solutions (i.e., these solutions are identified to be the best according to the secondary metric of crowding density).

In this case, however, it is expected that the inferior solutions recently explored are still capable of providing direction information. Let us consider the approach of JADE with archive. An external archive, $\mathbf{A}$, is created to store the parent individuals that are recently removed from the population because they are dominated by other solutions in the selection process; i.e., these individuals are removed from the population in the first two steps of the selection process, instead of the third step. Then, the mutant vector is generated as follows

$$\mathbf{v}_{i,g} = \mathbf{x}_{i,g} + F_i(\mathbf{x}^p_{\text{best},g} - \mathbf{x}_{i,g}) + F_i(\mathbf{x}_{r1,g} - \tilde{\mathbf{x}}_{r2,g}), \tag{6.6}$$

where $\tilde{\mathbf{x}}_{r2,g}$ is a vector randomly chosen from the union of the parent population and the archive, while $\mathbf{x}_{r1,g}$ and $\mathbf{x}^p_{\text{best},g}$ are randomly chosen from the parent population.

If the archive is not empty and $\tilde{\mathbf{x}}_{r2,g}$ is chosen from it, the difference between $\tilde{\mathbf{x}}_{r2,g}$ and $\mathbf{x}_{r1,g}$ carries the information about the direction towards the optimum. Note that this is different from the method NSDE proposed in [128], where direction information is obtained by comparing solutions in the current population that have different dominance ranks (thus, the direction information becomes unavailable if all solutions in the population become non-dominated). In addition, the operation in (6.6) is clearly different from many other MOEAs [58], [60], [61], [134] where

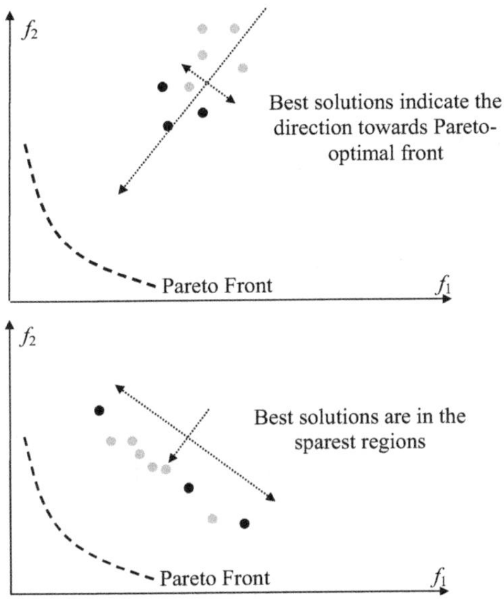

Fig. 6.1 An illustration of the relative positions of the best solutions (dark dots) and other solutions (gray dots) in the objective space. The unidirectional arrows show the progress direction towards the Pareto front, while the bidirectional arrows show the spread of the population orthogonal to the progress direction

an archive is maintained to store the best (i.e., non-dominated and well diversified) solutions obtained so far.

We consider rather simple archive operations to minimize the complexity. The archive is initialized to be empty. After each generation, we add to the archive those parent individuals that fail in the first two steps of the selection process. Then, if the archive size exceeds a threshold, say $2NP$, we randomly remove some solutions to keep the archive size at $2NP$.

6.4 Performance Comparison

In this section, we describe the performance matrices used for the evaluation of multi-objective optimization algorithms and present the comparison results based on the experimental studies on a set of twenty-five multi-objective test problems. These include the frequently used ZDT1 – ZDT4 and ZDT6 in [52], DTLZ1 – DTLZ4 and DTLZ7 in [136], and the three WKG functions [137] (WKG1, WFG8 and WFG9) that are believed to be hard problems because of the bias, discontinuity, and deception in the problems [138]. We also consider three test problems (QV, KUR, and SPH) previously studied in SPEA2 [60] and NSGA-II [35], and two recently proposed test problems SYMPART [139] and OKA2 [140].

We compare the performance of JADE with respect to NSGA-II and GDE3, as the former is an approach representative of the state-of-the-art, while the latter is a DE based algorithm that has shown competitive performance. JADE is also compared to MODE and SPEA based on the reported results in [15]. For fairness, we adopt the parameter settings in the original paper of NSGA-II [35] (i.e., $p_c = 0.9$, $p_m = 1/D$, $\eta_c = 20$ and $\eta_m = 20$) and GDE3 [34] (i.e., $F = 0.2$, and $CR = 0.2$). The parameters in JADE are set to be $c = 1/10$ and $p = 10\%$ in all simulations.

The number of generations is set to be 250 for all tested problems, except that 500 for SPH, WFG8 and WFG9, 2000 for QV and KUR, and 3000 for WFG1. The results are obtained based on 50 independent runs of each algorithm. In Table 6.3, we report both the final results and the intermediate results obtained after 60% of total generations.

6.4.1 Comparison Based on Conventional Performance Metrics

We first introduce several conventional performance metrics which have been used to compare different algorithms in the literature. One commonly used metric is the average Euclidean distance from the computed non-dominated front $\mathbf{Z}$ to the true Pareto front $\mathbf{Z}^*$,

$$d = \frac{1}{|\mathbf{Z}|} \sum_{\mathbf{z} \in \mathbf{Z}} \min\{\| \mathbf{z} - \mathbf{z}^* \|, \mathbf{z}^* \in \mathbf{Z}^*\}, \tag{6.7}$$

where $|\cdot|$ denotes the cardinality of a set and $\| \mathbf{z} - \mathbf{z}^* \|$ represents the Euclidean distance between two vectors $\mathbf{z}$ and $\mathbf{z}^*$ in the non-dominated front and the true Pareto front, respectively. Another one is the Pareto Front Approximation Error (PFAE) metric, which is defined as a tuple of three components in [15]. The PFAE metric is concerned with a calculated non-dominated solution, for which there exists at least one true Pareto optimal solution that is closer to it than to any other calculated non-dominated solutions. Denote $\mathbf{M}$ as the set of all calculated non-dominated solutions of this type. It is a subset of $\mathbf{z}$. Thus, we can group the true Pareto optimal solutions into $|\mathbf{M}|$ subsets according to their minimal distance to each of the solutions in $\mathbf{M}$. Denote D_i (i = 1, 2, ..., $|\mathbf{M}|$) as the set of the minimal distances related to each subset of true optimal solutions to their corresponding solution in $\mathbf{M}$. Then, the first component of PFAE is defined as follows,

$$\tilde{d}_{\min} := \text{median}(\min(D_i)), \tag{6.8}$$

which is the median of the minimal elements in all sets D_i, i = 1, 2, ..., $|\mathbf{M}|$. The $\tilde{d}_{\min}$ measures the closeness of the computed non-dominated front to the true Pareto front. The second component of PFAE is defined as follows:

$$\bar{d}_q := \frac{1}{|\mathbf{Z}_q^*|} \sum_{\mathbf{z}^* \in \mathbf{Z}_q^*} \min_{\mathbf{z} \in \mathbf{Z}} \| \mathbf{z} - \mathbf{z}^* \|, \tag{6.9}$$

where $\mathbf{Z}_q^*$ is a subset of $\mathbf{Z}^*$, containing a quarter of the Pareto optimal solutions that have the largest distances to $\mathbf{Z}$. The extra distance $\bar{d}_q - \tilde{d}_{\min}$ measures by how much the computed non-dominated front fails to reflect the whole picture of the true Pareto front. The cardinality $|\mathbf{M}|$ is defined as the third component of PFAE and it is considered less important than the first two components.

We first compare the different algorithms in terms of the average Euclidean distance from the calculated non-dominated solutions to the true Pareto optimal front. Table 6.1 summarizes the mean and the standard deviation of 50 independent runs of each function by each algorithm. The results of MODE and SPEA come from [15, Table 4.1]. We can see that JADE consistently obtains non-dominated fronts that are closer to the true Pareto front than all other algorithms. The obtained solutions of JADE are usually less than 10^{-5} from the true Pareto front, while those obtained by other solutions are greater than 10^{-3} from the true Pareto front in most cases.

We next perform a comparison of different algorithms based on the PFAE metric introduced above. Table 6.2 summarizes the mean and deviations of the three components of the PFAE metric obtained by each algorithm. It is clear that the results of JADE are much better than those obtained by other algorithms by comparing these statistical performance metrics. For example, the $\bar{d}_q$ of the computed non-dominated front using JADE is less than 0.01 in all cases except DZT6. This means that over 75% of the true Pareto solutions have an approximated computed solution that is within the 0.01 neighborhood. The $\bar{d}_q$ metric of JADE is degraded in the case of DZT6, but is still competitive to the best algorithm MODE in this case. In addition, the standard deviation of the results obtained by JADE is very small, indicating its consistent performance in different independent runs.

6.4.2 *Pareto-Compliant Performance Metrics*

In the past subsection, we have compared different algorithms based on conventional performance metrics. A potential drawback of the conventional metrics, however, is that they might not be *Pareto compliant* [141]. Define an *approximation set* as the set of non-dominated objective vectors that is an approximation to the true Pareto-optimal front. Roughly speaking, if a performance metric/indicator is not *Pareto compliant*, its means that an approximation set A may be assigned a better indicator value than another set B even if A is dominated or weakly dominated by B (i.e., every vector in A is weakly dominated or weakly dominated by at least one vector

Table 6.1 A comparison of the average Euclidean distance from the non-dominated front to the true Pareto front. The results of MODE and SPEA come from [15]

	JADE	NSGA-II	GDE3	MODE	SPEA
DZT1	**2.45E-6 (5.72E-7)**	1.50E-3 (2.27E-4)	1.08E-3 (2.67E-3)	0.003 (5E-4)	0.04 (0)
DZT2	**3.19E-6 (1.26E-6)**	1.56E-3 (2.82E-4)	4.00E-3 (7.13E-3)	0.003 (6E-4)	0.07 (0.02)
DZT3	**9.18E-6 (5.29E-6)**	6.75E-4 (1.14E-4)	3.72E-4 (9.15E-4)	0.006 (5E-4)	0.02 (0.01)
DZT4	**1.83E-3 (8.53E-3)**	6.01E-3 (4.75E-3)	4.85E-1 (2.49E-1)	0.723 (0.329)	4.28 (1.9)
DZT6	**1.81E-6 (1.08E-7)**	1.27E-2 (2.19E-3)	1.80E-6 (1.32E-7)	0.035 (0.023)	0.48 (0.15)

Table 6.2 A comparison of the PFAE metric. The results of MODE and SPEA come from [15]. From top to bottom, the three parts of the table correspond the three components of the PFAE metric: $\tilde{d}_{\min}$, $\bar{d}_q$, and |M|

	JADE	NSGA-II	GDE3	MODE	SPEA
DZT1	**2.34E-06 (6.05E-7)**	1.09E-03 (1.85E-4)	1.16E-03 (2.88E-3)	0.003 (0.0005)	0.036 (0.004)
DZT2	**3.09E-06 (1.17E-6)**	1.24E-03 (2.74E-4)	3.91E-03 (6.97E-3)	0.003 (0.0006)	0.066 (0.016)
DZT3	**2.18E-06 (3.48E-7)**	3.92E-04 (8.14E-5)	2.43E-04 (5.96E-4)	0.004 (0.0006)	0.012 (0.001)
DZT4	**1.85E-03 (9.18E-3)**	**5.41E-03 (3.62E-3)**	4.59E-01 (2.16E-1)	0.67 (0.25)	1.8 (1.65)
DZT6	**1.77E-06 (1.79E-7)**	1.24E-02 (2.00E-3)	**1.76E-06 (1.70E-7)**	0.01 (0.017)	0.43 (0.15)
DZT1	**6.84E-03 (6.05E-5)**	9.68E-03 (4.67E-4)	7.45E-03 (1.54E-3)	0.009 (0.0004)	0.046 (0.005)
DZT2	**6.80E-03 (5.41E-5)**	1.57E-02 (2.53E-2)	9.56E-03 (5.72E-3)	0.01 (0.0006)	0.093 (0.022)
DZT3	**8.64E-03 (6.80E-5)**	2.47E-02 (4.31E-2)	1.22E-02 (2.30E-2)	0.01 (0.001)	0.075 (0.021)
DZT4	**9.05E-03 (8.71E-3)**	9.67E-02 (1.01E-1)	4.82E-01 (2.34E-1)	0.75 (0.33)	4.07 (3.35)
DZT6	1.07E-01 (3.01E-5)	1.07E-01 (9.70E-4)	1.21E-01 (9.69E-2)	**0.02 (0.024)**	0.84 (0.23)
DZT1	**100 (0)**	98.84 (1.03)	99.90 (0.30)	98.2 (0.73)	46 (5)
DZT2	**100 (0)**	98.78 (1.06)	**100 (0)**	99.9 (0.18)	29 (5)
DZT3	**100 (0)**	98.98 (0.97)	99.70 (1.04)	85 (1.83)	49 (6)
DZT4	**99.88 (0.73)**	95.26 (8.67)	56.72 (16.94)	27 (11.3)	2 (2)
DZT5	**100 (0)**	94.52 (3.28)	**100 (0)**	50.8 (10.6)	4 (2)

in B). To the contrary, an indicator is Pareto compliant if there is no inconsistency between the indicator values and the Pareto dominance relationship for any two possible approximation sets A and B.

In this subsection, we follow the guidelines in [141] and use a set of *Pareto compliant* performance metrics such as dominance ranking, quality indicators and the empirical attainment surface for comparing the approximation sets obtained by different algorithms.

6.4.2.1 Dominance Ranking

Suppose the approximate sets generated by an evolutionary algorithm i after n independent runs are A_1^i, A_2^i, ..., A_n^i. If we consider the collection of approximate sets of all Q algorithms, then we can assign each approximate set a rank on the basis of dominance relationship, e.g., by counting the number of sets by which a specific approximation set is dominated (an approximation set A is said to be dominated by another set B if every objective vector in A is dominated by at least one vector in B). The lower the rank, the better the corresponding approximation set with respect to the entire collection. In this way, we obtain a set of ranks $\{r_1^i, r_2^i, \ldots, r_n^i\}$ for the approximation sets obtained by each algorithm. Then, a statistical rank test, such as the Mann-Whitney rank sum test [141], [142], can be applied to determine if one algorithm is significantly better than another algorithm in terms of the dominance ranking.

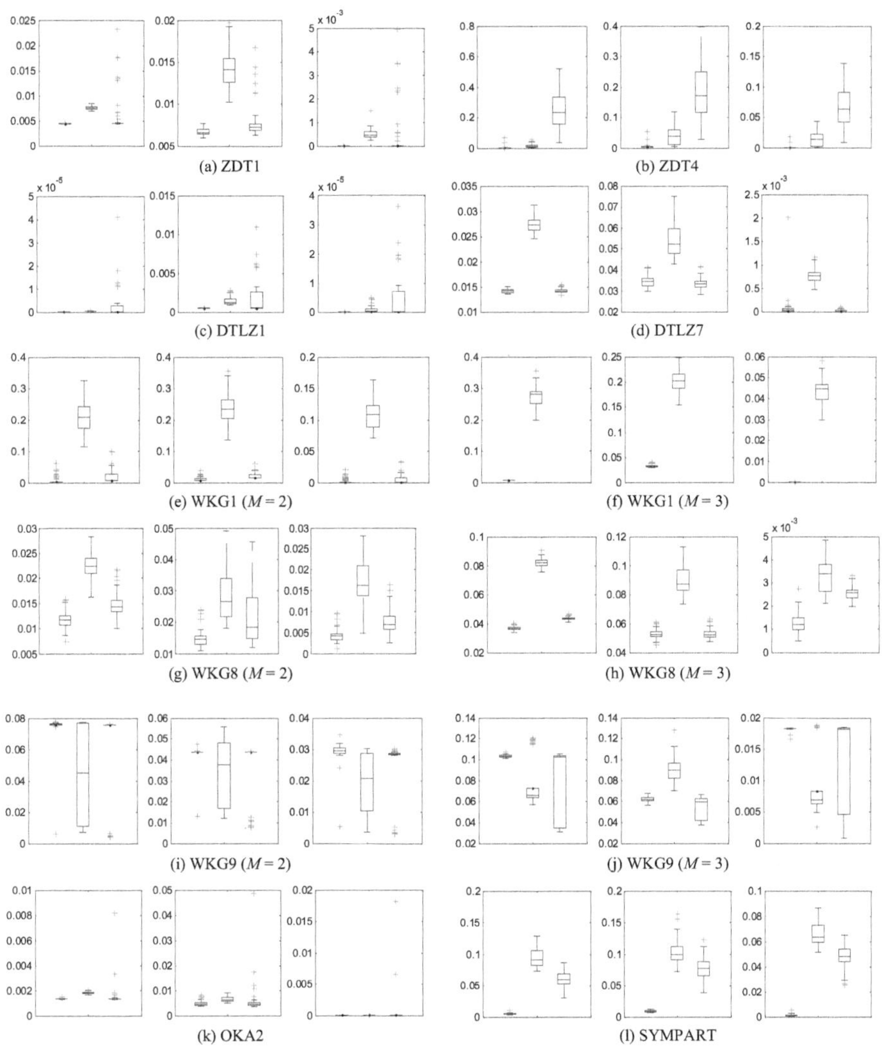

Fig. 6.2 A comparison of JADE, NSGA-II and GDE3 based on quality indicators. The three box and whisker plots in each sub-figure correspond to JADE, NSGA-II, and GDE3, respectively. The three sub-figures for each problem show the results of hyper-volume indicator I_H, unary additive epsilon indicator $I^1_{\varepsilon+}$, and unary $R2$ indicator I^1_{R2}, respectively

6.4.2.2 Quality Indicators

A quality indicator is a function that assigns a real value to any approximation set by means of some preference information. The difference between two indicator values reveals the superiority of one set over the other one in a certain aspect, even when the two sets are incomparable according to Pareto dominance. This type of information is usually considered secondary to Pareto dominance, and thus the indicators that are compliant to the Pareto relationship are more meaningful than Pareto non-compliant ones. In view of this, we consider the three Pareto compliant quality indicators recommended in [141]: (1) the *hyper-volume indicator* I_H^- that measures the difference between the hyper-volumes of the portion of the objective space that is weakly dominated by an approximation set or by a reference set, (2) the *unary additive epsilon indicator* $I_{\varepsilon+}^1$ that gives the minimum value ε that can be added to each vector in a reference set such that the resulting shifted set is weakly dominated by the approximate set of interest, and (3) the *unary R2 indicator* I_{R2}^1 that is used to assess an approximation set on the basis of a set of utility functions. All these indicators are to be minimized.

Table 6.3 A comparison of the intermediate results (after 60% of total generations) and final results obtained by different algorithms with respect to dominance ranking *DR* (Mann-Whitney rank sum test at a significant level 0.05), hyper-volume indicator I_H^-, unary additive epsilon indicator $I_{\varepsilon+}^1$, and unary R2 indicator I_{R2}^1 (Fisher-independent test at a significant level 0.05). The intermediate results are obtained after 60% of total number of generations. The symbol '$\triangle$' ('∇') means that JADE is significantly better (worse) than the other algorithm regarding to the dominance ranking or other indicators, respectively.

		JADE v.s. NSGA-II								JADE v.s. GDE3							
		Intermediate results				Final results				Intermediate results				Final results			
	M	*DR*	I_H^-	$I_{\varepsilon+}^1$	I_{R2}^1	*DR*	I_H^-	$I_{\varepsilon+}^1$	I_{R2}^1	*DR*	I_H^-	$I_{\varepsilon+}^1$	I_{R2}^1	*DR*	I_H^-	$I_{\varepsilon+}^1$	I_{R2}^1
ZDT1	2	-	$\triangle$	$\triangle$	$\triangle$	-	$\triangle$	$\triangle$	$\triangle$	-	$\triangle$	$\triangle$	$\triangle$	-	$\triangle$	$\triangle$	$\triangle$
ZDT2	2	$\triangle$	$\triangle$	$\triangle$	$\triangle$	-	$\triangle$	$\triangle$	$\triangle$	-	$\triangle$	$\triangle$	$\triangle$	-	$\triangle$	$\triangle$	$\triangle$
ZDT3	2	-	$\triangle$	$\triangle$	$\triangle$	-	$\triangle$	$\triangle$	$\triangle$	-	$\triangle$	$\triangle$	$\triangle$	-	$\triangle$	$\triangle$	$\triangle$
ZDT4	2	-	∇	∇	-	-	$\triangle$	$\triangle$	$\triangle$	-	-	∇	-	-	$\triangle$	$\triangle$	$\triangle$
ZDT6	2	-	$\triangle$	$\triangle$	$\triangle$	-	$\triangle$	$\triangle$	$\triangle$	-	$\triangle$	-	-	-	$\triangle$	$\triangle$	-
DTLZ1	3	-	$\triangle$	$\triangle$	$\triangle$	-	$\triangle$	$\triangle$	$\triangle$	-	$\triangle$	$\triangle$	$\triangle$	-	$\triangle$	$\triangle$	$\triangle$
DTLZ2	3	-	$\triangle$	$\triangle$	$\triangle$	-	$\triangle$	$\triangle$	$\triangle$	-	∇	-	∇	-	$\triangle$	$\triangle$	∇
DTLZ3	3	-	$\triangle$	$\triangle$	$\triangle$	-	$\triangle$	$\triangle$	$\triangle$	-	$\triangle$	$\triangle$	$\triangle$	-	$\triangle$	$\triangle$	$\triangle$
DTLZ4	3	-	$\triangle$	$\triangle$	$\triangle$	-	$\triangle$	$\triangle$	$\triangle$	-	∇	-	∇	-	$\triangle$	$\triangle$	∇
DTLZ7	3	-	$\triangle$	$\triangle$	$\triangle$	-	$\triangle$	$\triangle$	$\triangle$	-	-	-	∇	-	-	-	∇
WKG1	2	$\triangle$	$\triangle$	$\triangle$	$\triangle$	$\triangle$	$\triangle$	$\triangle$	$\triangle$	∇	∇	∇	∇	$\triangle$	$\triangle$	$\triangle$	$\triangle$
WKG1	3	$\triangle$	$\triangle$	$\triangle$	$\triangle$	$\triangle$	$\triangle$	$\triangle$	$\triangle$	∇	∇	∇	∇	$\triangle$	$\triangle$	$\triangle$	$\triangle$
WKG8	2	-	$\triangle$	$\triangle$	$\triangle$	-	$\triangle$	$\triangle$	$\triangle$	-	$\triangle$	$\triangle$	$\triangle$	-	$\triangle$	$\triangle$	$\triangle$
WKG8	3	-	$\triangle$	$\triangle$	$\triangle$	-	$\triangle$	$\triangle$	$\triangle$	-	$\triangle$	-	-	-	$\triangle$	-	$\triangle$
WKG9	2	∇	∇	∇	∇	∇	-	∇	∇	∇	∇	∇	∇	∇	∇	∇	∇
WKG9	3	-	∇	-	∇	-	∇	$\triangle$	∇	-	∇	∇	∇	-	∇	∇	∇
OKA2	2	-	$\triangle$	$\triangle$	-	-	$\triangle$	$\triangle$	$\triangle$	-	$\triangle$	$\triangle$	-	-	$\triangle$	-	-
SYMPART	2	$\triangle$	$\triangle$	$\triangle$	$\triangle$	$\triangle$	$\triangle$	$\triangle$	$\triangle$	-	$\triangle$	$\triangle$	$\triangle$	-	$\triangle$	$\triangle$	$\triangle$
QV	2	-	∇	∇	∇	-	-	∇	∇	-	-	-	$\triangle$	-	$\triangle$	$\triangle$	$\triangle$
KUR	2	$\triangle$	$\triangle$	$\triangle$	$\triangle$	$\triangle$	$\triangle$	$\triangle$	$\triangle$	-	$\triangle$	$\triangle$	$\triangle$	$\triangle$	$\triangle$	$\triangle$	$\triangle$
SPH	2	$\triangle$	$\triangle$	$\triangle$	$\triangle$	$\triangle$	$\triangle$	$\triangle$	$\triangle$	$\triangle$	$\triangle$	$\triangle$	$\triangle$	$\triangle$	$\triangle$	$\triangle$	$\triangle$
SPH	3	$\triangle$	$\triangle$	$\triangle$	$\triangle$	$\triangle$	$\triangle$	$\triangle$	$\triangle$	-	∇	-	∇	-	-	$\triangle$	∇

Similar to the dominance ranking, the indicator values obtained by each algorithm after n independent runs can be collected together and assigned rank values. Then, a statistical rank test, such as the Fisher-independent test [141], [142], can be applied to determine if one algorithm is significantly better than another algorithm in terms of the corresponding indicator.

6.4.2.3 Attainment Surface

The third approach to performance assessment is based on the multi-objective concept of *goal-attainment*: an objective vector is attained when it is weakly dominated by the approximation set obtained by an algorithm. The *k%-approximation surface* of the n approximation sets $A_1, A_2, \ldots, A_n$ obtained by an algorithm consists of the tightest objective vectors that have been attained in at least k percent of the n runs of an algorithm. That is, a k%-attainment surface roughly divides the objective space in two parts: the goals that have been attained and the goals that have not been attained with a frequency of at least k percent. The attainment surface is useful for visualizing an algorithm's performance.

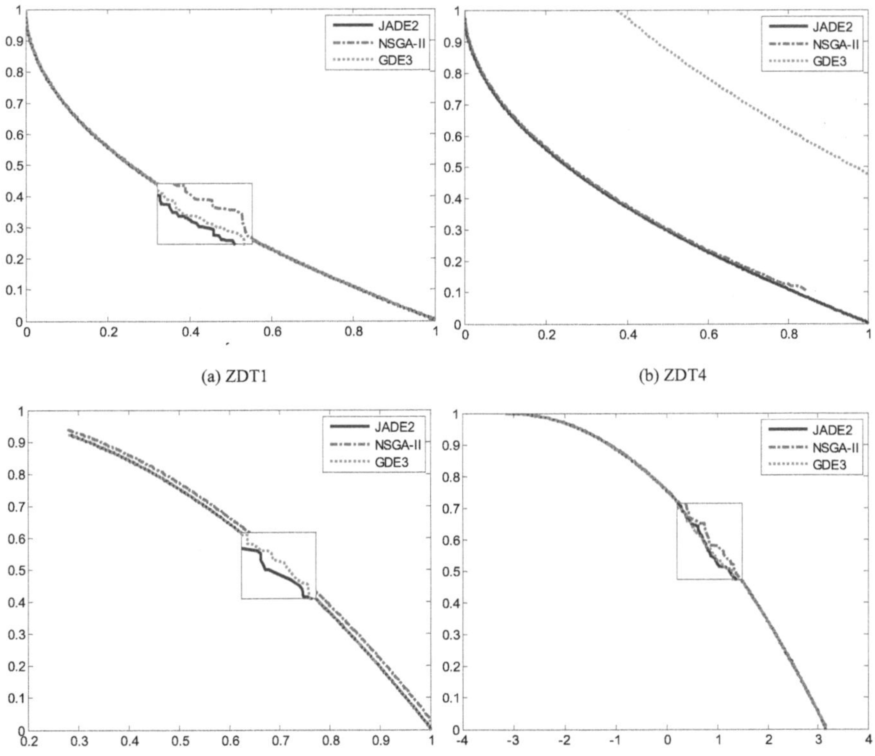

(a) ZDT1 (b) ZDT4

(c) ZDT6 (d OKA2

Fig. 6.3 The 50% attainment surfaces for two-objective problems. The results are obtained based on 50 independent runs of each algorithm

6.4.3 Experimental Results

Table 6.3 presents the outcomes of dominance ranking as well as the three quality indicators, hyper-volume indicator I_H^-, unary additive epsilon indicator $I_{\varepsilon+}^1$, and unary R2 indicator I_{R2}^1. In Figure 6.2, we also draw box and whisker plots for selected problems to have a clear comparison between JADE and other algorithms.

It is clear that JADE shows consistently better performance than NSGA-II for all problems studied in this chapter except WKG9 and QV. The dominance rank of JADE is significantly better than that of NSGA-II in the case of SYMPART, WKG1, KUR and SPH (In other cases except WKG9 and QV, the comparison results are statistically insignificant but a clear overall trend of Pareto dominance can still be observed as shown later in Figure 6.3). In terms of the three quality indicators, JADE shows significantly better performance for all studied problems except WKG9 and QV. In the case of WKG9, JADE may cause false convergence while NSGA-II has no difficulty of approaching the true optimal front. In the case of QV, both JADE and NSGA-II can easily find the true Pareto front but NSGA-II converges slightly faster.

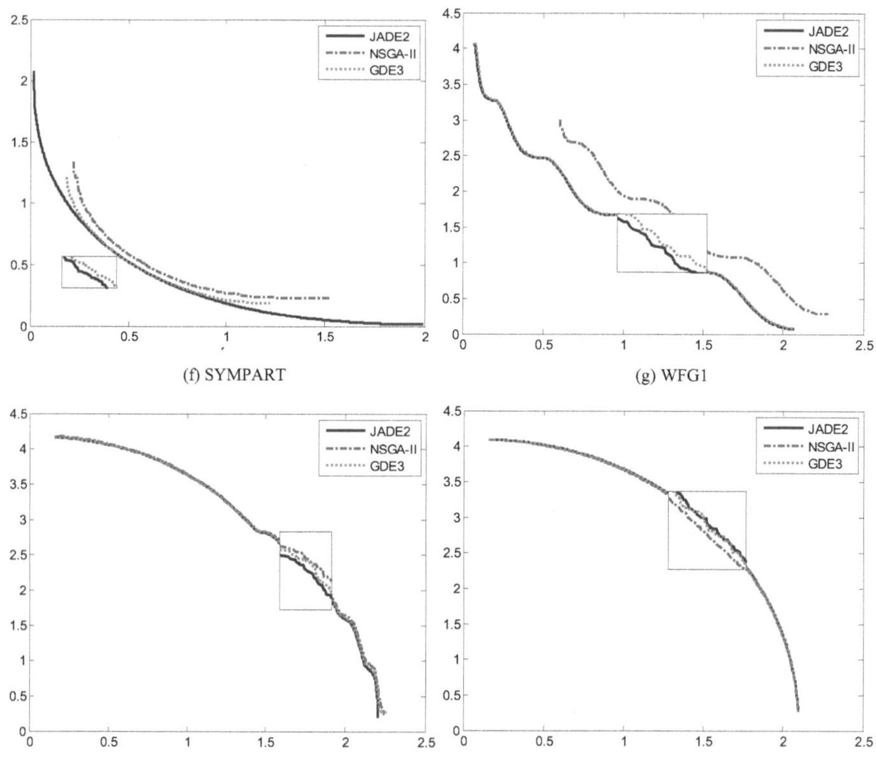

(f) SYMPART (g) WFG1

(h) WFG8 (i) WFG9

Fig. 6.4 The 50% attainment surfaces for two-objective problems. The results are obtained based on 50 independent runs of each algorithm

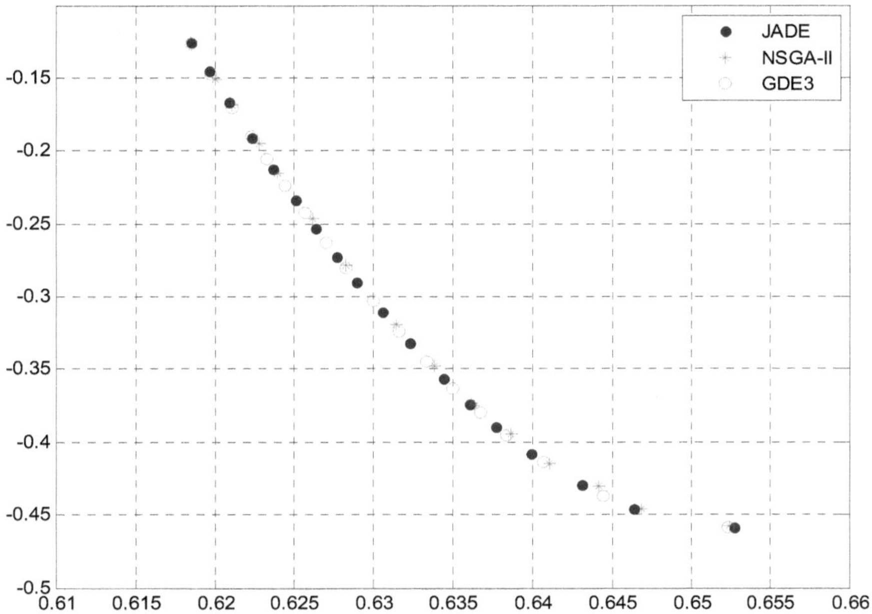

(a) ZDT3

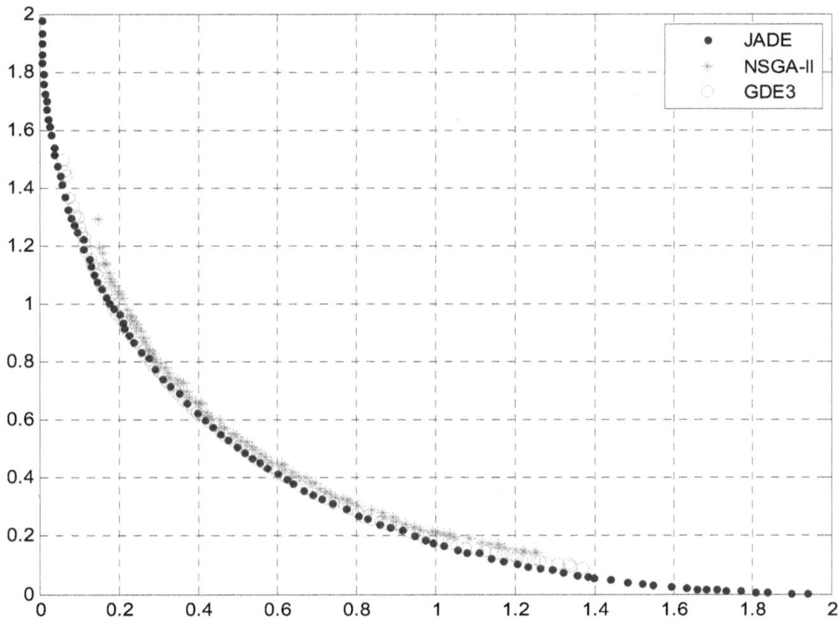

(b) SYMPART

Fig. 6.5 The best approximation sets (according to I_H^-) obtained by JADE, NSGA-II, and GDE3

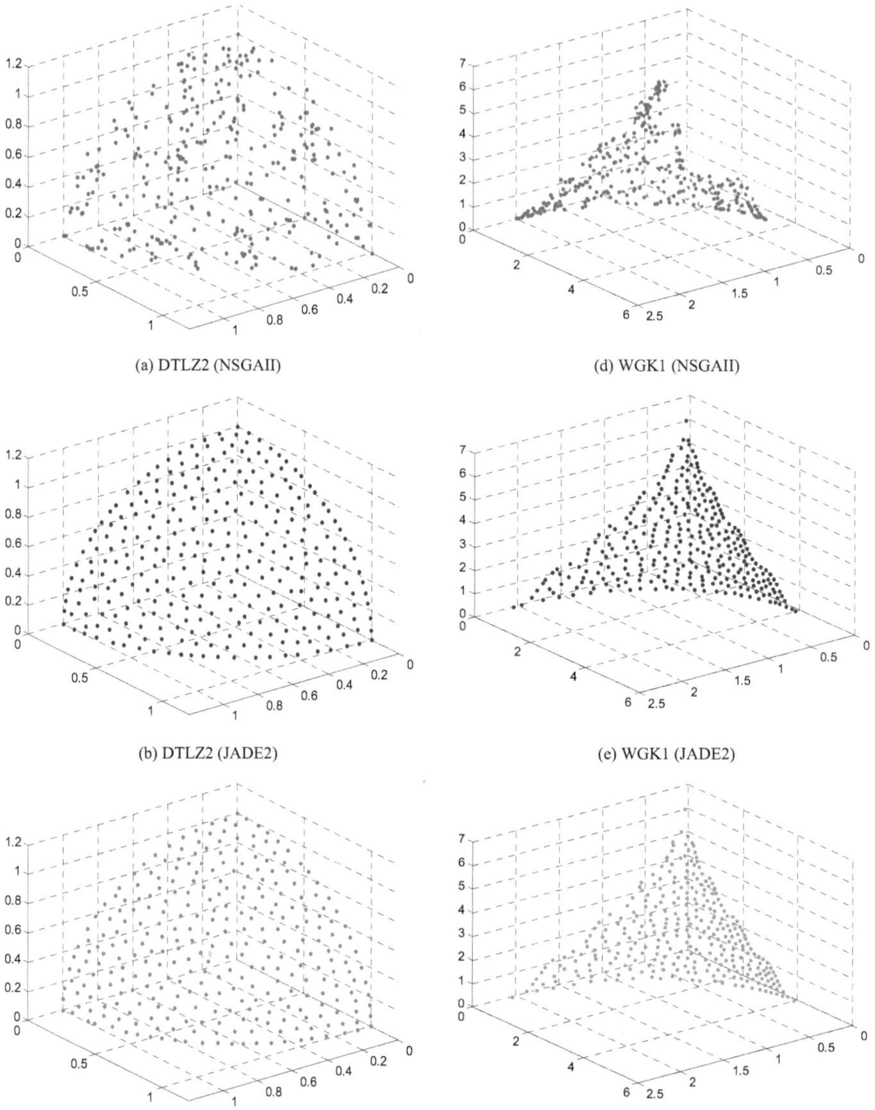

(a) DTLZ2 (NSGAII)

(d) WGK1 (NSGAII)

(b) DTLZ2 (JADE2)

(e) WGK1 (JADE2)

(c) DTLZ2 (GDE3)

(f) WGK1 (GDE3)

Fig. 6.6 The best approximation sets (according to I_H^-) obtained by JADE, NSGA-II, and GDE3

JADE shows on average better final performance than GDE3, especially when the hyper-volume indicator and unary additive epsilon indicator are concerned - JADE is better in 18 out of the 22 test problems. In terms of the *R*2 indicator, GDE3 is superior in six problems but is beaten by JADE in other 14 problems. Similar to the comparison with NSGA-II, however, JADE shows worse performance than GDE3 in the case of WKG9 because of more frequent false convergence.

Based on the performance comparison of intermediate and final results in Table 6.3, it is interesting that JADE gains more competitive advantages over GDE3 after a larger number of generations, especially in the case of ZDT4 ad WDK1. As explained before, the best solutions only direct the evolutionary search at the beginning of the optimization process, while the archived inferior solutions used in JADE are capable of providing direction information even at the later stage of the optimization when most or all solutions in the population become non-dominated. Thus, JADE is able to keep a relatively fast convergence and outperform GDE3 after a larger number of generations.

To visualize the final solutions obtained by different algorithms, we plot in Figure 6.3 – Figure 6.6 the 50% attainment surfaces for some two-objective problems and the best approximation sets (according to the I_H^- indictor) calculated based on the results of 50 independent runs. These graphs are consistent with the statistic results summarized in Table 6.3 and Figure 6.2, which indicate a generally better performance of JADE than NSGA-II in terms of both the overall trend of Pareto dominance (as shown in Figure 6.3) and the spread of final results (as shown in Figure 6.5). The spread of the final results of GDE3 and JADE is usually incomparable, mainly because they adopt the similar product distance to estimate the crowding density. However, the comparison in Figure 6.3 indicates that in most cases JADE is able to attain solutions which are clearly better than or at least competitive to those of GDE3.

6.5 Summary

We have proposed a new differential evolution algorithm JADE for multi-objective optimization by controlling the parameters in an adaptive manner and utilizing the direction information provided by an external archive. Adaptive parameter control avoids users' prior knowledge of the fitness characteristics of optimization problems and therefore avoids tedious parameter tuning. The external archive is used to store recently explored inferior solutions whose difference with the current population indicates promising directions towards the optimum and thus can be used to speed up the evolutionary search. JADE shows better results than NSGA-II for 18 out of 22 test problems in terms of the overall trend of Pareto-dominance and three quality indicators. It outperforms GDE3 as well, especially in terms of the hypervolume indicator and unary additive epsilon indicator. In addition, JADE gains more competitive advantages over GDE3 after a larger number of generations, indicating its relatively fast convergence rate at a later stage of evolutionary search because of the direction information supplied by archived solutions.

A future research direction is to study the proposed algorithm in solving optimization problems with many (>3) objectives and to investigate the effect of parameter α of the fairness measure (used in calculating crowding distances) on achieving diversified solutions. It will be meaningful to compare JADE to other latest algorithms such as [143] that have shown promising results in solving many-objective optimization problems.

Chapter 7
Application to Winner Determination Problems in Combinatorial Auctions

The crucial idea behind DE is the scheme of generating mutant vectors using the weighted difference of other vectors randomly chosen from the population. This mutation operation relies on arithmetic operations (add, subtract, multiply, etc.) rather than general data manipulation operations (sort, swap, permute, etc.). It cannot be directly extended to the discrete combinatorial space.

In this chapter, we apply parameter adaptive differential evolution to a seminal combinatorial optimization problem of winner determination in Combinatorial Auctions (CAs). To adapt JADE to this problem, we use a rank-based representation scheme that produces only feasible solutions and a regeneration operation that constricts the search space. It is shown that JADE compares favorably to the classic DE algorithm, a local stochastic search algorithm Casanova, and a genetic algorithm based approach SGA.

7.1 Introduction

In differential evolution, mutant vectors are generated by weighted differences of other vectors randomly chosen from the population. It relies on arithmetic operations (add, subtract, multiply, etc.) rather than general data manipulation operations (sort, swap, permute, etc.). It cannot be directly extended to the discrete combinatorial space. Therefore, DE algorithms have been traditionally developed for optimization problems where the search space is continuous, while the extensions to discrete combinatorial optimization problems are relatively small. Therefore, DE algorithms have been traditionally developed for optimization problems where the search space is continuous. It is extended to discrete combinatorial optimization problems mainly by rounding real-valued parameters to the nearest discrete values in the differential mutation operation [144], [145] or proposing new mutation strategies which are not of conventional DE-style (i.e., the randomness introduced in the mutation does not rely on the difference between parent vectors) [146], [147]. The performance of DE, however, is not efficient for combinatorial problems whose discrete parameter values have no physical meaning and thus do not measure quantity. In addition, it

J. Zhang and A.C. Sanderson: Adaptive Differential Evolution, ALO 1, pp. 115–125.
springerlink.com

shows that for some classic combinatorial problems DE fails to recognize identical/equivalent solutions or generating a high proportion of infeasible solutions that need to be repaired [2].

A seminal combinatorial optimization problem is winner determination in combinatorial auctions [148]. In combinatorial auctions, agents are allowed to bid on a combination of items, and express any complementarities directly and are used in such environments with strong synergistic relationships between goods. For example, combinatorial auctions have been used for resource allocation in sensor management [149], [150], supply chain management [151] and computer grids [152]. However, the winner determination in combinatorial auctions, i.e., finding out the optimal categorization of bids as either "winning" or "losing" is an NP-hard problem [153]. Various approaches have been suggested in the literature to solve this problem, including exact optimization algorithms like CPLEX [154] [155] and CABOB [156] and approximate optimization techniques Casanova [157] and SGA [158]. For many realistic distributions, CPLEX and CABOB perform extremely well with average running time being less than a second for problem sizes with thousands of bids. However, for certain complex bid distributions, neither CPLEX nor CABOB's real-time performance is adequate in the environments with strict real-time constraints. For example, for certain bid distributions, the complete algorithms took extremely long completion times even for medium sized problems (see [158] for details). Researchers have used heuristic approaches to solve the "hard" bid distributions. Two of the most recent approaches that have been shown to outperform others are Casanova and SGA. Casanova, a local stochastic search procedure, is an approximate winner determination algorithm, based on Novelty+. SGA is a standard genetic algorithm that uses a modified representation scheme to enhance its performance.

JADE has been tested on continuous search spaces and its applicability to combinatorial problems has not been tested. To apply JADE to the combinatorial auction problems, we consider a rank-based representation scheme that transforms the discrete combinatorial problem into continuous domain and only produces feasible solutions to the problem. Also, we propose a regeneration operator to represent equivalent solutions into a unique form and to constrict the search space that is previously enlarged due to the continuous representation of the discrete search space. Furthermore, to improve the convergence performance, we initialize the algorithm with a set of solutions that are randomly generated in the neighborhood of a seeded high-quality solution. The resultant methodology, the seeded JADE with regeneration operation, is shown to consistently achieve better solutions than Casanova and SGA for two combinatorial auction problems that are toughest for the exact winner determination algorithms [154].

The rest of this chapter is organized as follows. Section 7.2 introduces the winner determination problem and the scheme for representing this discrete combinatorial problem into continuous domain. Section 7.3 proposes regeneration and seeding operations to improve the performance of JADE in optimizing the discrete combinatorial problem. In Section 7.4, we compare the performance of JADE with other approximate algorithms. The final section presents our conclusions and directions for future work.

7.2 Problem Description and Current Approaches

This section describes the winner determination problem and briefly introduces the best solution approaches in the literature.

7.2.1 *Winner Determination in Combinatorial Auctions*

Assume that the auctioneer has a set of m goods or items $G = \{g_1, g_2, \ldots, g_m\}$ to sell, and a set of n bids $B = \{b_1, b_2, \ldots, b_n\}$. Bidders offer bids of the form $B_i = < b_i, p_i >$, where b_i is the bundle of goods and p_i is the price the bidder is willing to pay. The winner determination problem is to find an allocation of goods that maximizes the overall utility, given the constraint that each good can be sold to no more than one bid. Formally, this problem can be formulated as [153]:

$$\max \sum_{j=1}^{n} p_j x_j \quad \text{s.t.} \sum_{j \in S(i)} x_j \leq 1, \forall i \in \{1, 2, \ldots, m\} \tag{7.1}$$

where $S(i)$ is the set of bids that contain good i, and x_j is a binary variable that equals 1 if bid j is accepted to the final allocation and 0 otherwise. A solution is feasible if and only if each good appears in at most one accepted bid.

7.2.2 *Current Approaches*

In the literature, the various algorithms proposed for the winner determination problem can be classified into two categories:

1. Exact winner determination: the algorithm guarantees the convergence to global optimum. Examples are CPLEX [154] and CABOB [156].
2. Approximate winner determination: the algorithm does not provide optimality guarantees. Examples are Casanova [157] and SGA [158].

The most successful approach for exact winner determination is solving it as an integer-programming problem using standard integer programming (IP) software like CPLEX [154]. CPLEX has been shown to perform remarkably well on a wide-range of problem distributions (see [155] for details).

Another example of exact winner determination procedure is CABOB [156], which uses a specialized depth first search that branches on bids. CABOB constructs a *bidgraph* with each bid in the set of bids Bas the vertex of the graph and edges between vertices only when they have items in common. CABOB then uses a separate depth-first search (DFS) to find the optimal solution. By selecting an initial bid as being in a possible allocation, then remaining bids with items that do not overlap this bid can also be in the allocation and are considered for allocation in a particular search path of the DFS. CABOB bounds the DFS by pruning paths that have already been dominated.

CABOB's performance has been found to be marginally better than CPLEX for certain hard to solve problems [156]. However, when the problem sizes are large and when the bid-distributions are not simple, both CABOB and CPLEX fail to scale up well [158].

In these cases, approximate winner determination procedures are useful. The most prominent approach in this genre has been Casanova, a local stochastic search procedure [157]. Casanova starts with an empty allocation and bids are chosen from the unsatisfied bids randomly and added to the allocation vector. The probability that a bid is chosen is determined by its normalized score, calculated as the bid price divided by the product of the number of goods the bid consumes, the bid's age, and the number of steps since it has last been chosen, to be added to the allocation vector. Recently, a genetic algorithm based approach, SGA [158], is proposed by utilizing a novel representation method that restricts evolutionary search only to feasible solutions. SGA performs better than Casanova for problems with a large number of bids.

7.3 Utilization of Domain Knowledge

As introduced in Chapter 2.5, the performance of an algorithm can be improved by utilizing problem-specific knowledge. In this section, we introduce a new solution representation scheme based on the characteristic of the winner determination problems and propose two new problem-specific operations, regeneration and seeding, to improve the performance of JADE.

7.3.1 Representation Scheme for Discrete Optimization

Techniques such as differential evolution and particle swarm optimization [159], which function by manipulating the coordinates of solution vectors by arithmetic operations (add, subtract, multiply, divide, etc.), are not directly applicable to discrete optimization problems. Recent developments to address this drawback involve converting the search space to a continuous one based on some transformations [160], [161], [162]. A popular approach among them is proposed by Kennedy and Eberhart [160], which, instead of searching over the binary space, operates on the continuous space of probabilities that a binary variable takes value 0 or 1. This approach is applicable to the combinatorial auction problem as the decision variables x_j in (7.1) take binary values. However, an important drawback is that it causes a high proportion of infeasible solutions where multiple bids containing a common item are marked as winning and therefore lead to poor convergence performance [158].

To avoid the problem of infeasible solutions, we derive from the idea in [158] which produces only feasible solutions in discrete space using a rank-based representation scheme. In view of the high efficiency of differential evolution in continuous optimization, we extend this representation scheme to transform the problem

Table 7.1 A problem instance. The accept list and reject list are updated according the acceptation or rejection of bids

A solution vector (0.91, 0.90, 0.13, 0.81)
The corresponding rank vector: (4, 3, 1, 2)

Rank	bid	Items in the bid	State	Accept list	Reject list
1	3	a, b, c	Win	{3}	ϕ
2	4	a, d, f	Lose	{3}	{4}
3	2	d, e	Win	{3, 2}	{4}
4	1	b, d, f	Lose	{3, 2}	{4, 1}

into a continuous search space. A regeneration operation will be proposed later to constrict the complexity of the problem by ensuring that all equivalent solutions in the continuous space are represented in a unique form.

In the rank-based representation scheme, each solution is represented by a vector $\mathbf{x}$ of length n whose elements take values between 0 and 1; i.e., $\mathbf{x} = (x_1, x_2, \ldots, x_n)$, with $x_i \in [0, 1]$. The vector $\mathbf{x}$ can be decoded into an ordered *accept list* and an ordered *reject list*, based on which the solution can be obtained. The decoding process is given as follows:

1. Generate a rank vector $\mathbf{z} = (z_1, z_2, \ldots, z_n)$ with $z_i = k$ if x_i is the k-th smallest element in $\mathbf{x}$; i.e., the i-th bid is considered to have a rank k.
2. Initialize the accept list with the lowest ranking bid, and set the reject list to be empty.
3. Consider the bid with the next lowest rank. Add it to the end of the accept list if it does not create an infeasible solution; otherwise add it to the end of reject list. Proceed to the bid with the next lowest rank.
4. Continue till the bid with the highest rank is analyzed.

For example, consider a scenario with 4 bids and 6 items (a, b, ..., f) as shown in Table 7.1. A solution vector (0.91, 0.90, 0.13, 0.81) is decoded in the following manner. According to the sorting result, the four elements of the vector are assigned ranks 4, 3, 1, and 2, respectively. The initial accept list is set to be {3}, as bid 3 has the lowest rank (rank 1). The next lowest ranking bid is bid 4 (rank 2). However, bid 4 has item a, which is also present in bid 3. Since bid 3 has already been labeled as winning, bid 4 is labeled as losing and added to the reject list. The procedure is repeated until the status of the highest-ranking bid (bid 1 in this case) is determined. At the end, the vector (0.91, 0.90, 0.13, 0.81) corresponds to a solution where bids 3 and 2 are marked as winning (i.e., accepted to the final allocation) and bids 4 and 1 marked as "losing".

7.3.2 Regeneration Operation

JADE, similar to other differential evolution algorithms, is originally designed for continuous optimization problems. In the last section, we have transformed the winner determination problem into a continuous domain by the rank-based representation scheme. The method, however, causes a problem to JADE for the following reason. The rank-based scheme is unable to detect identical or equivalent solutions, because what matters in the scheme is the rank (instead of the value) of the element in a solution vector. For example, two vectors (0.3, 0.5, 0.6, 0.7) and (0.1, 0.2, 0.7, 0.9) lead to the same rank vector (1, 2, 3, 4) and thus represent an identical solution. However, their difference, i.e., (0.2, 0.3, -0.1, -0.2), is not close to zero. As a result, JADE (especially its mutation operator) may not function efficiently because the difference between two solutions in the decision space (e.g., the Euclidean distance) has a weak relation to the difference in the objective space (i.e., the difference of the fitness values).

To avoid the above problem, we propose a regeneration operation to represent all *equivalent* solutions that lead to the same accept list and reject list in a unique form. Taking a solution vector $\mathbf{u} = (0.91, 0.90, 0.13, 0.81)$ in Table 7.1 as example, we can describe the regeneration operation as follows:

1. Concatenate the accept list and the reject list into a vector. This vector shows the sequence of bids getting accepted or rejected. Denote this *sequence* vector as $\mathbf{s} = (s_1, s_2, \ldots, s_n)$. For the example in Table 7.1, we have $\mathbf{s} = (3, 2, 4, 1)$.
2. Generate a new vector

$$\mathbf{u}' = (u_1', u_2', \ldots, u_n'), \text{ with } u_i' = j/n. \tag{7.2}$$

   if bid j is the i-th bid in the sequence vector (i.e., $s_j = i$). We have $\mathbf{u}' = (0.75, 0.50, 0.25, 1.00)$.
3. Add a small random disturbance to u', and we obtain

$$\mathbf{u}'' = \mathbf{u}' - \text{rand}(0, 1/n), \tag{7.3}$$

   where rand(0, $1/n$) is a uniform random number on the interval (0, $1/n$]. We have $\mathbf{u}'' = (0.67, 0.42, 0.17, 0.92)$ if, for example, rand(0, $1/n$) = 0.07.
4. Replace $\mathbf{u}$ with $\mathbf{u}''$.

It is easy to show that both $\mathbf{u}'$ and $\mathbf{u}''$are equivalent to $\mathbf{u}$. Indeed, we can obtain a unique $\mathbf{u}'$ for all equivalent solution vectors that are different in floating-point values but lead to the same accept and reject lists (and therefore the same sequence vector $\mathbf{s}$). It is worth noting that two vectors are considered as equivalent if and only if their respective accept/reject lists contain the same elements and the elements are listed in the same order. Take the problem in Table 7.1 as example. The vector (0.91, 0.90, 0.13, 0.81) leads to an accept list $\{3, 2\}$ and a reject list $\{4, 1\}$. Now, consider another vector (0.91, 0.13, 0.90, 0.81) which leads to an accept list $\{2, 3\}$ and a reject list $\{4, 1\}$. In both cases, bids 2 and 3 win, and bids 1 and 4 lose. However, the two vectors are not equivalent in the sense that the former (the latter) indicates a higher priority of bid 2 (bid 3) and thus more likely produces offspring that again include bid 2 (bid 3) in the accept list.

By representing all equivalent solution in a unique form $\mathbf{u}'$, the problem complexity, which is inevitably enlarged due to the continuous representation of the discrete search space, can be greatly constricted. In (7.3), we add a small random disturbance to $\mathbf{u}'$, because otherwise the difference of individual vectors only take values that are a multiple of $1/n$, causing degraded performance in the mutation operation. The random disturbance is small enough not to affect the rank of the elements (and thus $\mathbf{u}'$ and $\mathbf{u}''$are equivalent), so the problem's complexity is not changed due to (7.3).

7.3.3 Seeding JADE

If a high quality solution is available before the optimization, we are able to boost the performance of an algorithm by initializing solutions in the neighborhood of the high quality solution. The following solutions can be considered as the candidates of high quality solutions:

1. The solution obtained by a less-efficient yet fast approximate winner determination algorithm.
2. The solution obtained by calculating a quick lower bound for the winner determination problem using linear programming . For example, for CAT distributions and weighted random distributions [156] , a linear programming solution is usually close to the optimum.
3. The optimal or high quality solution previously obtained in an incremental winner determination process. Assume that the auctioneer has calculated an optimal or high quality solution for the bids received. If certain bids are retracted or new bids are added (e.g., in the sensor management scenario [149]), the auctioneer can use the previous solution as a high quality solution to the new problem.

In this chapter, we consider a simple heuristic method that finds a high quality solution $\mathbf{x}^{\mathrm{hq}}$ based on the unit price of a bid (i.e., the price of a bid divided by the number of items in it). The basic reasoning is that the higher the unit price of a bid, the higher the priority of this bid being considered in the auction. To be specific, the i-th element x_i^{hq} of $\mathbf{x}^{\mathrm{hq}}$ is assigned a value j/n, if bid i has the j-th highest unit price among all the n bids.

We next randomly generate a set of seed solutions in the neighborhood of the high-quality solution obtained above. The k-th seed solution $\mathbf{x}^k = (x_1^k, x_2^k, \ldots, x_n^k)$ is obtained in the following manner:

$$x_i^k = x_i^{\mathrm{hq}} + \mathrm{rand}_i(0, k/n), \text{ for } i = 1, 2, \ldots, n. \tag{7.4}$$

It is clear that the larger the k, the more disturbance is added to the elements of $\mathbf{x}^{\mathrm{hq}}$. In the case of $k = 1$, $\mathbf{x}^1$ is equivalent to $\mathbf{x}^{\mathrm{hq}}$ because the random disturbance is always less than $1/n$, the minimum difference between the elements of $\mathbf{x}^{\mathrm{hq}}$, and thus does not affect the rank of the elements.

7.4 Performance Comparison

In this section, we compare JADE with other prominent approximate winner determination algorithms, Casanova and SGA. We also include the classic differential evolution algorithm DE/rand/1/bin [1] in the comparison to demonstrate the benefit of the features introduced in JADE.

7.4.1 Experimental Setting

We perform two sets of experiments on the uniform and bounded bid distributions, as they are shown to be the toughest for the exact winner determination algorithms [154].

1. Uniform (n, m, λ) bid distribution which consists of n bids, each with λ items chosen without replacement from the m items. Price is chosen randomly from a uniform distribution on [0, 1]. For our experiments, we use $m = n/10$ and $\lambda = 5$, as in [156].
2. Bounded (n, m, λ^1, λ^2) bid distribution where the problem consists of n bids. The number of items λ is randomly chosen between a lower bound λ^1 and an upper bound λ^2. The price is chosen from a uniform distribution on [0, λ]. For our experiments, we use $m = n/10$ and $\lambda^1 = 5$ and $\lambda^2 = 10$, as in [156].

The parameter values of JADE are selected as $c = 1/10$, and $p = 5\%$, as suggested in Section 4.4.4. In DE/rand/1/bin, we set $F = 0.5$ and $CR = 0.9$, as recommended in [1], [9], [10]. For fair comparison, the regeneration and seeding operations proposed in JADE is also applied to DE/rand/1/bin. In both algorithms, the number of generations is set to be 500, and the population size is $NP = n/2$, which is much smaller than the value $3n \sim 10n$ as usually suggested in the literature of DE [1], [2]. The small population size speeds up the convergence rate at the expense of the exploration capability of the algorithm. However, as shown below, the final performance JADE is still significantly better than other algorithms.

For Casanova, we follow the parameters suggested in the original paper [157]. However, we find that changing the walk probability to 0.4 gives better results and hence $w_p = 0.4$ is used. The SGA parameters are chosen from [158]: the population size is set to 6000 if $n \leq 1400$ and 6500 otherwise; the tour size is equal to 2; and the crossover and mutation probabilities equal 0.995 and 0.02, respectively. Note that the population size of SGA is much greater than that of JADE.

We compare the four algorithms on large problems, varying the number of bids from 1000 to 2200 bids. In each case, the results reported below are calculated based on the evaluation of each algorithm on 30 problem instances randomly generated according to the uniform or bounded bid distribution.

7.4.2 Comparison Results

Figure 7.1 illustrates the performance comparison between JADE and other algorithms. JADE shows consistently better performance than Casanova for the problems with both uniform and bounded bid distribution. The benefit is slightly increasing (from 2$\sim$3% to 4$\sim$5%) as the problem size goes up from 1000 to 2200.

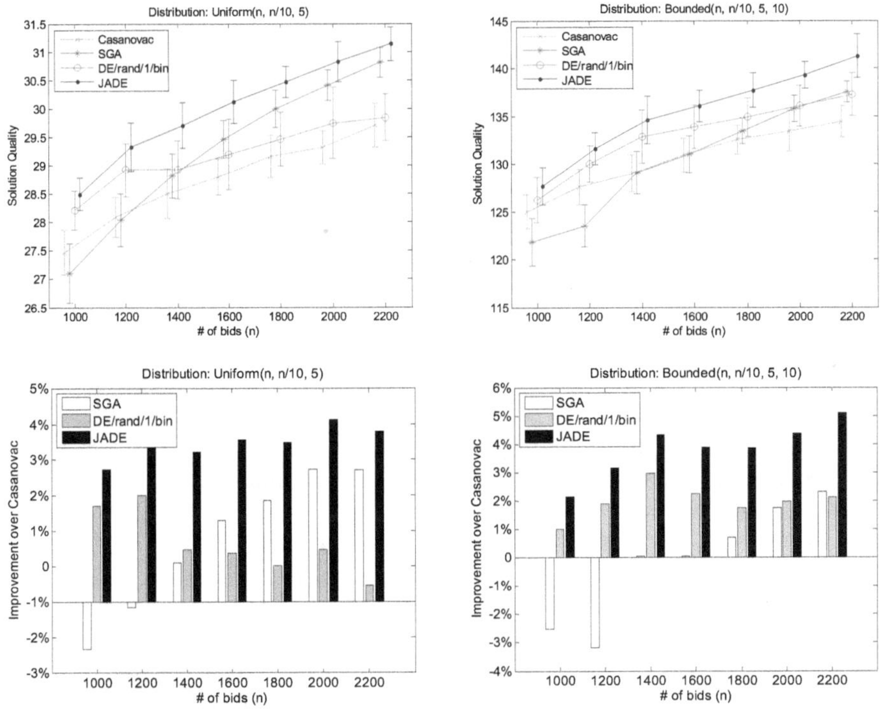

Fig. 7.1 A performance comparison among JADE, Casanova, SGA, and DE/rand/1/bin. The top two sub-figures show the mean and standard deviation of the best solutions obtained by each algorithm. For clarity, the error bar plots are slightly shifted for each algorithm in the sub-figures. The bottom two sub-figures show the improvement of JADE, SGA and DE/rand/1/bin over Casanova

In addition, there is no increasing trend in the variation of results, indicating similar robustness of JADE and Casanova for problems of different sizes.

JADE performs better than SGA as well. As shown in Figure 7.1, SGA is worse than Casanova in the case of small problems but is better than Casanova when the problem size becomes greater than 1200 or 1400 in the case of uniform or bounded bid distributions, respectively. JADE outperforms Casanova in all the conducted experiments.[1] The performance of SGA gets closer to that of JADE as the problem size increases: while it trails JADE by a large margin (about 5%) for small problems, SGA becomes 1% or 3% worse than JADE for large problem in the case of the uniform or bounded bid distributions, respectively. However, there is no trend that SGA could beat JADE for even larger problems. It is worth noting that these comparison results are obtained with JADE using a much smaller population than SGA. In addition, considerable extra experimentations have been conducted to select the optimal set of parameters used in SGA for these combinatorial auction problems.

[1] This is the case even when the problem size is as low as 200. For smaller problems, both JADE and Casanova find solutions competitive to the results of exact winner determination algorithms.

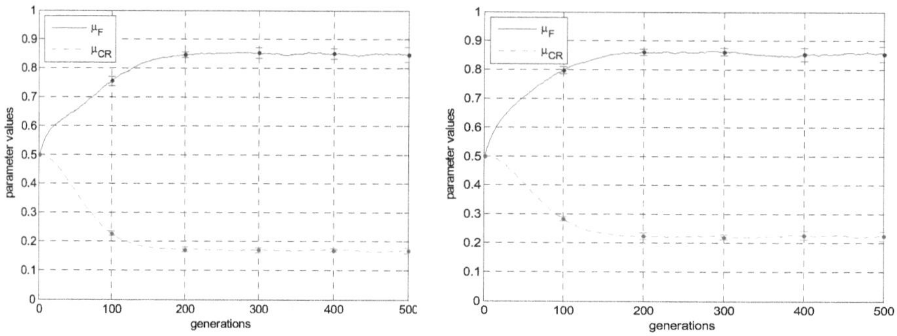

Fig. 7.2 The evolutions of μ_F and μ_{CR} in the optimization of problems with uniform (left plot) or bounded (right plot) bid distributions. The number of bid is 2200

As a comparison, parameter selection for JADE is very simple owing to the ability of parameter adaptation.

As JADE gets better performance than SGA, it is interesting to investigate if its benefit comes solely from the underlying differential evolution strategy or stems from its other unique features as well. The comparison of JADE and SGA with the classic differential evolution algorithm DE/rand/1/bin provides some information in this aspect. As shown in Figure 7.1, DE/rand/1/bin achieves much better results than SGA for small problems but does not continue performing efficiently for large problems. In all cases, however, DE/rand/1/bin is clearly worse than JADE, indicating the effectiveness of the new features (parameter adaptation and DE/current-to-*p*best mutation strategy) introduced in JADE.

The effect of parameter adaptation in JADE is illustrated in Figure 7.2. We show the evolution process of μ_F and μ_{CR} with mean curves and error bars. The error bars show a small variance of μ_F and μ_{CR} at different stages of the optimization, indicating a clear evolution trend of μ_F and μ_{CR} over 30 independent runs for both bid distributions. As the evolution process is automatic, JADE is able to find appropriate

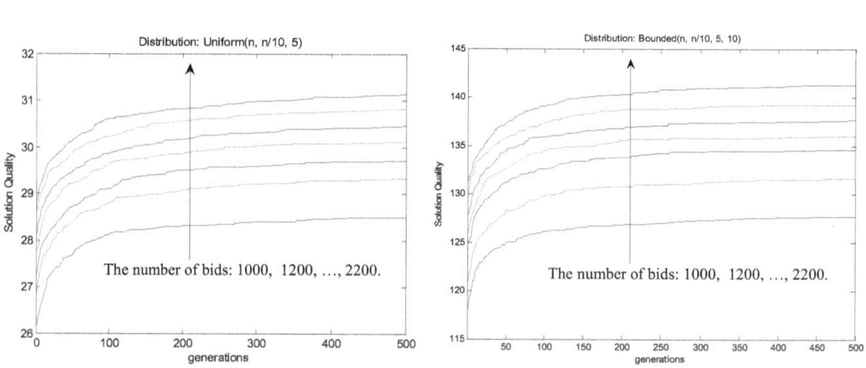

Fig. 7.3 The convergence graph of JADE (the mean of the results in 30 independent runs)

control parameter values without a user's prior knowledge of parameter settings or interaction during the evolutionary search.

It is also interesting to investigate the scalability of JADE for winner determination problems. Illustrated in Figure 7.3 is the convergence graph of JADE where the mean of the 30 random problems are plotted at each generation. It shows that JADE usually reaches a performance plateau after 100 to 200 generations, regardless of the problem size. In addition, we observe that JADE achieves better results than Casanova after only 20 $\sim$ 50 generations by comparing the convergence graphs to the results of Casanova in Figure 7.1.

7.5 Summary

Differential evolution has been successfully applied to a wide range of continuous optimization problems. In this chapter, we apply the adaptive differential evolution algorithm JADE to the discrete winner determination problem in combinatorial auctions. We transform the problem into the continuous domain by a rank-based representation scheme, which however inevitably enlarges the search space. Thus, we further propose a regeneration operation to represent all equivalent solutions in a unique form and thus significantly reduce the problem complexity. JADE outperforms both the local stochastic approach Casanova and the genetic algorithm based approach SGA, indicating its efficiency in the domain of discrete optimization.

Although the proposed methodology for representing a discrete optimization problem in a continuous domain is applicable directly only to combinatorial auctions, we believe that this technique can be extended to a large set of discrete optimization problems that involve the sorting or permutation of elements.

Chapter 8
Application to Flight Planning in Air Traffic Control Systems

Multi-objective optimization arises commonly in real-world applications such as automation controls, financial and scientific applications. In such an application, the outcome may be directly linked to cost, profit and/or many other inevitably conflicting criteria that have heavy impacts on the overall performance. An air traffic control system is such an example. In flight planning, air traffic controllers need to take into account different performance metrics or optimization objectives such as total flight distances, departure and arrival delays, system-level airspace congestion, stakeholders' preferences, etc.

In this chapter, we formulate the flight planning in air traffic control systems as a multi-objective optimization problem. While the original problem usually involves hundreds and thousands of decision variables, domain knowledge can be leveraged to remarkably reduce the dimension and therefore the complexity of the problem. The parameter adaptive differential evolution algorithm is then applied to the optimization problem and its performance is compared to other state-of-the-art algorithm.

8.1 Introduction

It is projected that the air traffic demands will double or triple in the next 20 years and thus worsen the current U.S. and European national air traffic systems that are currently operating at the edge of their capabilities. The imbalance between limited capacity and emerging heavy demand imposes an urgent need to review air traffic control systems in the near future. Additional capacity can be achieved through the introduction of new operational methods supported by advanced navigation capabilities and information technology. However, it is equally important to incorporate innovative optimized planning solutions to fully utilize the available capacity based on the information of predicted demand, the system-level behavior of air traffic system and the domain knowledge for the selection of optimization algorithms. It is

J. Zhang and A.C. Sanderson: Adaptive Differential Evolution, ALO 1, pp. 127–134.
springerlink.com

expected that a flight planning approach can maintain or improve system efficiency and provide flexibility to air traffic controllers.

As proposed in [163], the flight planning can be considered under a scalable enterprise framework of multi-stakeholder, multi-objective model-based planning and optimization in support of future air traffic flow management. The key objective is to not only ensure that flight operator objectives are balanced against the National Airspace System (NAS) performance, but also to ensure an equitable consideration of multiple stakeholder (airline operators) needs in this complex dynamic system.

For the simplicity of mathematical analysis as well as practical implementations, the multi-stakeholder, multi-objective model-based planning and optimization can be addressed based on an intelligent evaluation at the strategic and flight route levels. At the strategic level, one focuses on the separations among flights in the airspace to improve system performance. At the flight route level, one is concerned about identifying an optimal portfolio of flight routes within a planning horizon that trades off a reduction in miles flown and a reduction in congestion. Interested readers are referred to [163], [164] for the description and approach of strategic-level optimization. In this rest of this chapter, we address the flight-route level optimization. As is clear below, the flight route planning and optimization is by nature a multi-objective optimization problem.

8.2 Problem Formulation

The objective of automatic flight planning in the NAS is to produce a set of air traffic operations over a certain period (e.g. an entire day) that is compatible with the available capacity. The flight planning at an air traffic control system typically starts at midnight, while stakeholders (aircraft dispatchers) submit flight requests throughout the day. All scheduled carriers submit a flight plan, which contains one or multiple flight routes, for each flight prior to departure. Given a set of routes for each flight, we aim at identifying an optimal portfolio of flight routes (i.e., selecting an appropriate route for each flight) within a planning horizon that can best balance the reduction of total flight distance and the reduction of system-level congestion.

As shown in Figure 8.1, let us consider a simple scenario where each of the three flights has two alternate routes to their respective destinations. A flight route portfolio can be represented as a vector (x_1, x_2, x_3) with $x_i = 1$ or 2 denoting the first or second route is selected for the i-th flight, respectively. There are 8 possible portfolios in this simple situation. By evaluating the performance of all portfolios, we are able to find the solution that achieves the best tradeoff of the total flight distance and the system-level congestion. However, as the number of feasible portfolios increases expoentially fast and there are usually thousands of flights on a daily basis, it may quickly become impossible to explore all solutions to find the best tradeoff solutions that can balance the two competing objectives. An evolutionary algorithm based approach is capable of tackling this problem and achieving high-quality solutions within an acceptable period of time.

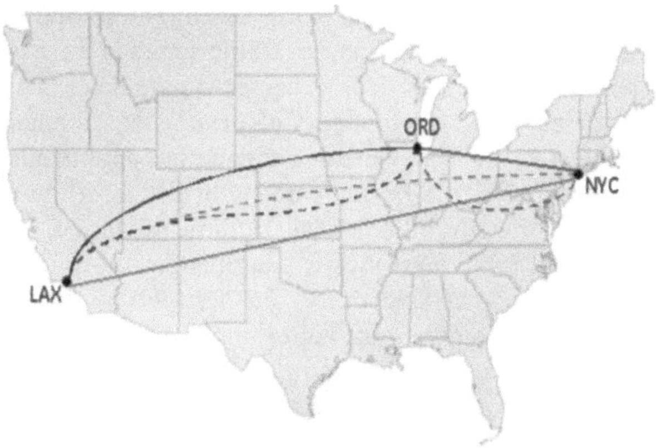

Fig. 8.1 An example of multiple flight routes scenario. (Courtesy of Subbu et al. [164])

The total flight distance can be easily calculated given a certain portfolio. That is, given a portfolio $\mathbf{x} = (x_1, x_2, \ldots, x_n)$ for a system with n flights, the total flight distance f_1 is give by

$$f_1(\mathbf{x}) = \sum_{i=1}^{n} d(i, x(i)), \tag{8.1}$$

where $d(i, j)$ denotes the length of the j-th route of the i-th flight. The system-level congestion, on the other hand, is a rather complex function of the selected portfolio because of the usually nonlinear correlation among flights in air traffic systems. An air traffic simulator is useful to evaluate the system level congestion but is usually very expensive. For convenience, we consider a simple metric to represent the congestion level, which ignores the correlation among flights. Denote $h(i, j)$ as the number of hotspot sectors that the j-th route of the i-th flight crosses, where the hotspot sectors are predefined before the flight planning and refer to the sectors that have a high flight density as a function of time. As a simple metric, the number of hotspot sectors can be used as an approximation to the system-level congestion. That is, the system level congestion f_2 is given as follows:

$$f_2(\mathbf{x}) = \sum_{i=1}^{N} h(i, x(i)). \tag{8.2}$$

The analysis below focuses on the optimization of the two objectives in (8.1) and (8.2). It is worth noting that our analysis is still applicable when a more accurate model (say, a simulation-based model), instead of (8.2), is used to evaluate the system-level congestion.

As the flight planning problem has been formulated into a multi-objective optimization framework, we apply the parameter adaptive DE algorithm JADE to solve it in the rest of the chapter. The benefit of adopting a parameter adaptive evolutionary algorithm such as JADE is twofold. First, evolutionary algorithms are capable of efficiently searching the whole solution space and generating a set of Pareto-optimal solutions in a single run of optimization. Second, as the air traffic varies in different air-traffic systems and at different time periods, it is expected a parameter adaptive algorithm can perform better than other evolutionary algorithms by automatically adapting its control parameters to the characteristics of different optimization landscapes. The proposed algorithm JADE will be compared with a state-of-the-art evolutionary algorithm NSGA-II [35] for several practical air traffic scenarios.

Once a set of Pareto-optimal flight route portfolios is identified, air traffic controllers can conduct decision-making or down-selection to one portfolio for deployment according to additional practical considerations. For example, a certain airline may prefer mileage savings, while another may prefer to reduce the number of route intersections with hotspot sectors. These preferences may be utilized in the Pareto-optimal down-selection process, which is however beyond the scope of our analysis in this chapter.

8.3 Utilization of Domain Knowledge

Similar to [164], let us take the historical NAS data of Chicago Center (ZAU) as an example. It involves a total of 77 sectors and 586 east coast to west coast and other intermediate en route flight traffic. The information of several flights is illustrated in Table 8.1. It is clear that the flights can be categorized into two classes according to the comparability of their two route options. Flights 1 – 3 belong to one class because their two routes are not comparable; i.e., if one route is better than the other one in terms of the flight distance, it is worse in terms of the congestion level.

To the contrary, the two routes of flights 4 and 5 are directly comparable. For example, the second route of flight 4 is shorter than the first one and is as good as (no worse than) the latter in terms of congestion. In this case, the second route of flight 4 is a better choice than the first one, regardless of the route selection of any other flights. According to the definition in (8.1) and (8.2), the second route can be safely assigned to flight 4 (i.e., $x(4) = 2$) to achieve an optimal portfolio. Similarly, we have $x(5) = 1$. Thus, we can conduct a pre-filtering procedure to immediately select a route for a flight if this route is identical to or dominates the other one. These flights can be ignored in the subsequent optimization process from the perspective

Table 8.1 An illustration of the flight information

Flights	Route 1		Route 2	
	$d(i,1)$	$h(i,1)$	$d(i,2)$	$h(i,2)$
$i = 1$	1897	3	2742	3
$i = 2$	917	9	720	4
$i = 3$	914	7	977	6
$i = 4$	2841	4	2832	4
$i = 5$	1573	1	1836	2

Table 8.2 Three flight planning scenarios

	Number of flights	
	Before pre-filtering	After pre-filtering
Scenario 1	586	60
Scenario 2	417	36
Scenario 3	1003	103

of joint optimization of flight distance and congestion level. It will greatly simplify the complexity of the optimization problem. As is shown in the example in [164], a route can be immediately assigned to 526 out of the 586 flights, as it is identical to or dominates the other route. This reduces the optimization problem to the selection of the best combination of routes for only 60 flights.

8.4 Simulation

We consider three different air traffic scenarios to validate the performance of JADE. The flight planning in these scenarios has recently been analyzed and solved by a state-of-the-art multi-objective evolutionary algorithm NSGA-II in [164]. As shown in Table 8.2, the number of flights involved in the optimization can be significantly reduced by the pre-filtering process introduced in the last section. The optimization is thus conducted only on the remaining flights.

For fairness, we adopt the parameter settings in the original paper of NSGA-II [35] (i.e., $p_c = 0.9$, $p_m = 1/D$, $\eta_c = 20$ and $\eta_m = 20$). The parameters in JADE are set to be $c = 1/10$ and $p = 5\%$ in all simulations. The population size is set to be 100 and the number of generations is set to be 250 for all tested problems. All the results are obtained based on 50 independent runs of each algorithm.

The performances of JADE and NSGA-II are evaluated and compared in two methods. First, we follow the method in Chapter 6.4 and calculate the three quality indicators, hyper-volume indicator I_H^-, unary additive epsilon indicator $I_{\varepsilon+}^1$, and unary R2 indicator I_{R2}^1. In Figure 8.2, JADE shows clearly better performance than NSGA-II in terms of the three performance metrics in all the three scenarios. As is clear from Figure 8.3, we note that these two algorithms are incomparable in terms of dominance ranking: the non-dominated front obtained by NSGA-II overlaps that obtained by JADE; though the solutions of JADE are more diversified and cover a larger region. For example, the solutions obtained by JADE can reduce the congestion level down to 71 (hotspot intersections), 76, and 169 in the three scenarios, respectively, while those obtained by NSGA-II down to 78, 79, and 180, respectively.

We next compare the performance of JADE and NSGA-II with the baseline metrics that are calculated for the portfolios where all flights use their first route options (preferred routes) or all use their second options (alternative routes). That is, we

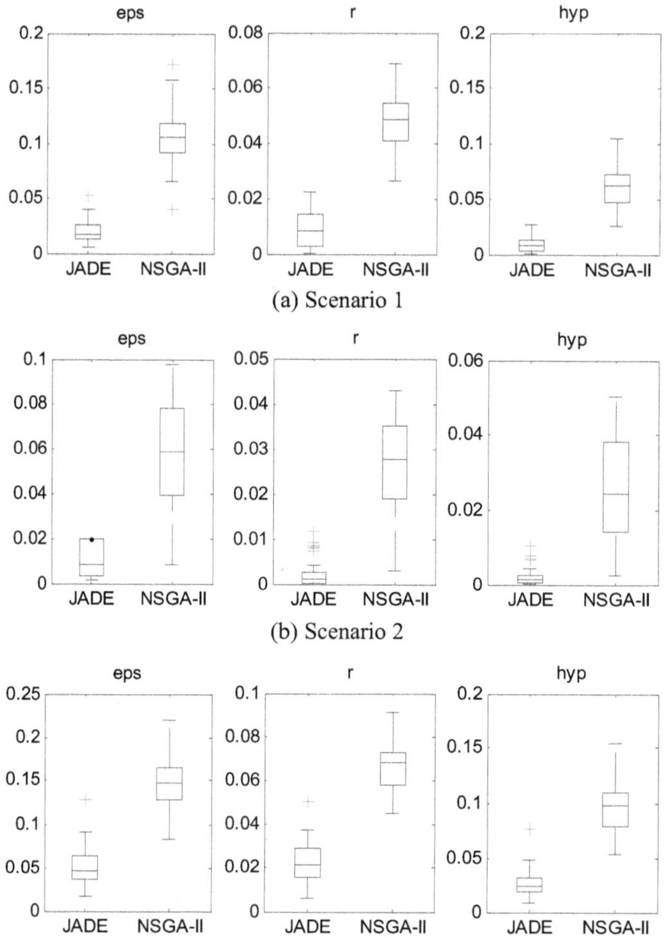

(a) Scenario 1

(b) Scenario 2

(c) Scenario 3

Fig. 8.2 A comparison of quality indicators. The three plots in each scenario show the results of hyper-volume indicator I_H^-, unary additive epsilon indicator $I_{\varepsilon+}^1$, and unary $R2$ indicator I_{R2}^1, respectively

compare the two key metrics (flight distance and congestion level) for four scenarios: the optimized results of JADE, the optimized results of NSGA-II, the result of all option 1 (each flight takes its first option), and the result of all option 2 (each flight takes its second option). The results are summarized in Figure 8.4. Take the better one between all option 1 and all option 2 as the baseline. The congestion (the number of hotspot intersections) can be reduced by NSGA-II by 17.0%, 19.4% and 15.4% (or by JADE by 24.5%. 22.4%, and 21.0%) in the three scenarios, respectively. The distance reduction is relatively small: 2.6%, 8.2%, and 3.0% by NSGA-II, and 3.4%, 8.5%, and 3.6% by JADE.

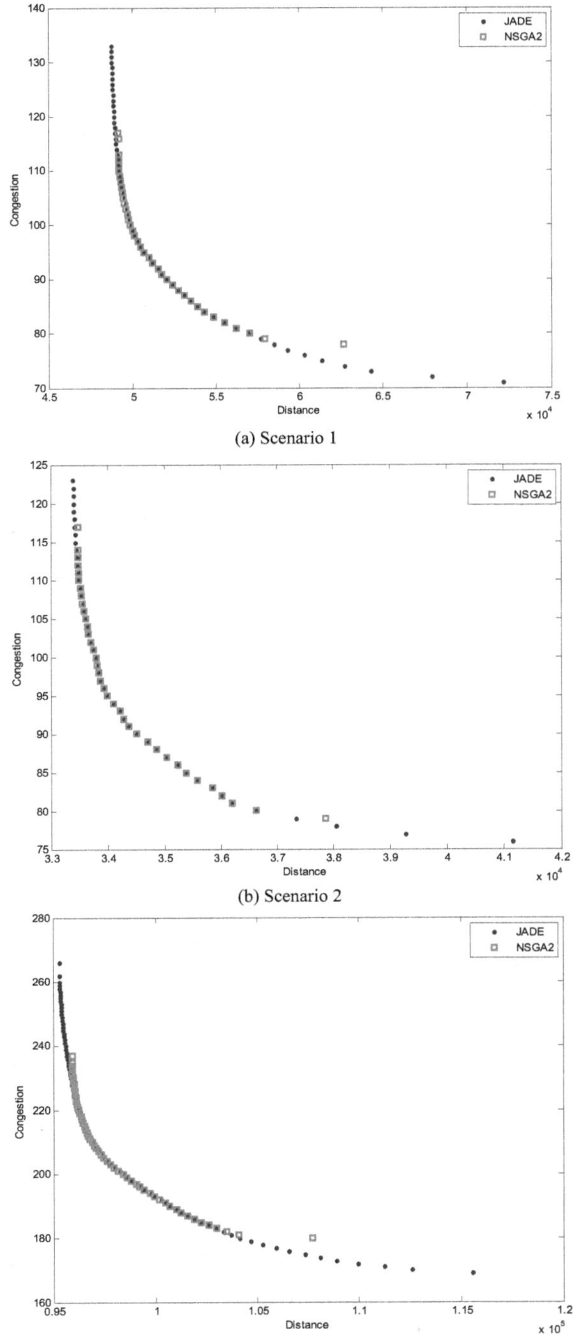

(a) Scenario 1

(b) Scenario 2

(c) Scenario 3

Fig. 8.3 The 50% attainment surfaces in the flight route optimization. The results are obtained based on 50 independent runs of each algorithm

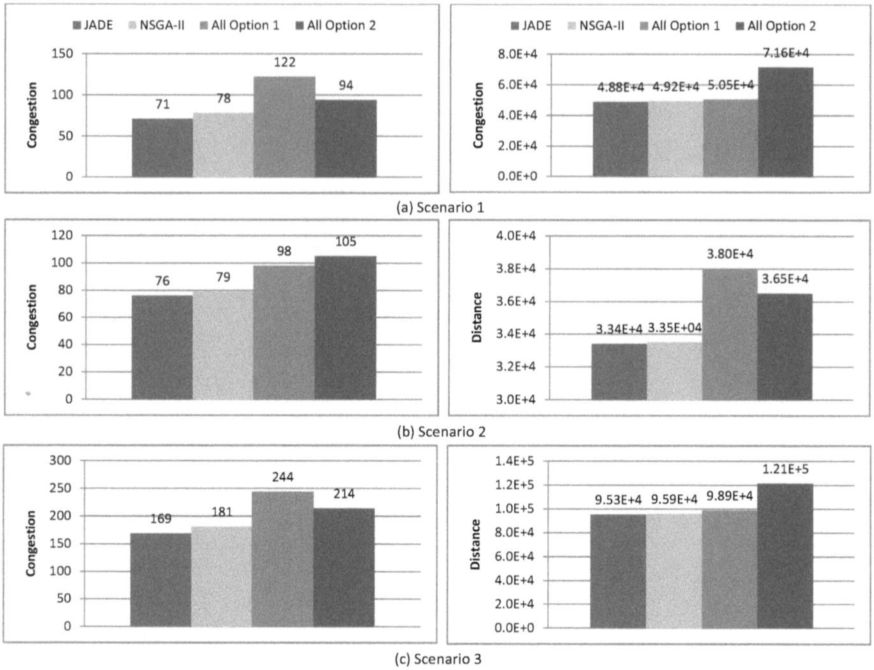

Fig. 8.4 A comparison of the optimal results obtained by JADE and NSGA-II and the results of all option 1 and all option 2

8.5 Summary

In this chapter, we have formulated the flight planning in air traffic control systems as a multi-objective optimization problem. Multi-objective evolutionary algorithms are then applied to solve the resultant problem after a pre-filtering process where domain knowledge is utilized to reduce the problem complexity. It shows that, comparing with two baseline solutions, the state-of-the-art genetic algorithm NSGA-II can significantly reduce the system congestion level by 17.3% on average and reduce the total flight distance by 4.6%, while the proposed parameter adaptive differential evolution JADE performs better by reducing the two metrics by 22.6% and 5.2%, respectively. Compared to NSGA-II, the solutions obtained by JADE are more diversified and provide more flexibility for the decision maker to balance between the congestion level and the total flight distance.

Chapter 9
Application to the TPM Optimization in Credit Decision Making

Statistical transition probability matrices (TPMs), which indicate the likelihood of obligor's credit rating migrating from one state to another over a given time horizon, have been used in various credit decision-making applications. Empirical TPMs calculated from historical data generally do not satisfy desired properties. An alternative method is to formulate the problem into an optimization framework [165], i.e., to find an optimized TPM that, when projected into the future based on Markov and time-homogeneity assumptions, can minimize the discrepancy from empirical TPMs. The desired properties can be explicitly modeled as the constraints of the optimization problem.

This TPM optimization problem is high dimensional, non-convex, and non-separable and is not effectively solved by nonlinear programming methods. It however can be well addressed by the proposed parameter adaptive DE algorithm where domain knowledge can be efficiently utilized to improve performance. In this chapter, we apply the proposed algorithm to this TPM optimization problem and compare its performance to a set of competitive algorithms.

9.1 Introduction

Credit ratings, which are the measures of obligors' creditworthiness, are issued to corporations by credit rating agencies, such as Standard & Poor's (S&P), Moody's or Fitch Ratings on a periodic basis [166]. These standard rating agencies use a variety of metrics, including the per capita income, GDP growth, inflation, external debt, level of economic development, and default history to determine the credit rating of a particular corporation (See [166] for details). The credit ratings are assigned as letter designations to corporations to indicate their creditworthiness. For example, if S&P assigns an AAA rating to a corporation, then S&P considers the corporation to be a highly reliable borrower. On the other hand, an S&P rating of D (the state of default) indicates that the corporation has a high probability of defaulting on financial obligations.

The likelihood of all obligor's credit rating migration, or the statistical transition probability from one credit rating to another one, has been used in various

J. Zhang and A.C. Sanderson: Adaptive Differential Evolution, ALO 1, pp. 135–145.
springerlink.com

high-dollar-value credit decision-making applications ranging from the pricing of financial instruments, loan evaluation, portfolio risk analysis, and economic capital assessment. Based on historical data of credit ratings, a rectangular matrix of transition probabilities between all possible credit ratings (including the state of default) can be constructed. Indeed, empirical TPMs (ETPMs) have been regularly published by rating agencies for many years [167] and users can use them either directly or after applying a certain smoothing technique.

A typical empirical TPM is shown in Table 9.1, where the i-th row j-th column element of the matrix represents the probability of migration from the i-th credit rating to the j-th rating (AAA, AA, ..., C, or Default). The need for TPM data smoothing stems from the fact that empirical matrices are strongly affected by idiosyncratic historical events, sparse data for plausible future scenarios, and vintage effects caused by the temporal location of the underlying data relative to the credit cycle. Practitioners usually calculate one-year empirical TPMs by a discrete cohort approach and iteratively apply the one-year matrix for the analysis over a multi-year horizon. However, the problem of one-year empirical TPMs (e.g., the TPM in Table 9.1) is that they usually lack desired properties such as structural stability and satisfaction of default probability constraints.

- Structural stability: the TPM should show the natural tendency of obligors to maintain status quo and to have a greater probability of migrating to a nearer credit rating than a farther one. In other words, it is expected that in a TPM matrix the diagonal value for any particular rating is greater than off-diagonal values and the off-diagonal values decrease monotonically with the distance to the diagonal term.
- Default probability constraints: the probability of default (the transition from a regular rating to default) is expected to satisfy monotonic properties: the default probability increases over time for each credit rating, and in each year a higher rating tends to have a lower probability of default than a lower rating.

Table 9.1 An example of one-year empirical transition probability matrix. The i-th row j-th column element represents the probability of migration from the i-th credit rating to the j-th rating (AAA, AA, ..., C, or Default). Bold cells indicate structural or default constraint violations

Credit ratings	AAA	AA	A	BBB	BB	B	CCC	C	Default
AAA	85.98%	11.72%	1.76%	0.43%	0.12%	0.00%	0.00%	0.00%	0.00%
AA	1.79%	83.32%	12.07%	2.21%	0.31%	0.00%	0.00%	***0.06%***	0.24%
A	0.17%	3.58%	84.05%	10.31%	1.03%	0.20%	0.15%	***0.17%***	0.34%
BBB	0.33%	0.56%	5.06%	85.39%	7.44%	0.40%	0.32%	0.18%	***0.32%***
BB	0.01%	0.04%	0.42%	3.70%	82.60%	6.34%	3.83%	0.91%	2.15%
B	0.02%	0.07%	0.13%	0.20%	5.67%	71.85%	14.14%	2.32%	5.59%
CCC	0.00%	0.03%	***0.66%***	0.17%	0.54%	5.01%	69.90%	5.74%	17.95%
C	0.00%	0.00%	***0.32%***	0.00%	0.90%	1.09%	11.15%	46.86%	39.68%
Default	0.00%	0.00%	0.00%	0.00%	0.00%	0.00%	0.00%	0.00%	100.00%

An alternative approach is to pose the computing and smoothing of a one-year TPM as a constrained optimization problem. The objective function to be minimized is an error function that calculates the discrepancy between the predicted transition matrices and the empirical data over the required time horizon. All the required structural properties of the one-year TPM are captured in the form of constraints. The problem however is complex due to the following reasons:

- The objective function is highly nonlinear due to the nonlinear nature of usual error functions and the matrix exponential operation involved in calculating the later-year transition probability matrices.
- The problem dimension, proportional to the number of variables in the TPM, is of the order of hundreds considering that there may be tens of credit ratings;
- The optimized one-year TPM is expected to satisfy structural and default properties.

Due to the above difficulties, a traditional nonlinear programming method is not efficient to find the global optimal solution. This motivates a consideration of the parameter adaptive differential evolution, as it has shown promising results on a set of nonlinear, non-convex benchmark problems with no requirement of tedious parameter tuning. It is expected that the large number of constraints involved in this TPM optimization problem will pose a new challenge to our research. In solving this problem, the proposed algorithm JADE is to be compared with both traditional nonlinear programming solver as the *fmincon* in MATLAB [88] and other evolutionary algorithms as introduced in Chapter 4. The performance of these algorithms will be evaluated in terms of both the time complexity and the quality of the final solutions obtained (the discrepancy of the optimized TPM to the empirical TPM matrices over a certain time horizon).

9.2 Problem Formulation

In this section, we formulate the problem into an optimization framework as proposed in [165]. The thereafter development and performance analysis are mainly based on the materials in an earlier study of [38].

The industry standard for estimating discrete-time credit transition probability matrices is the cohort approach [168]. This approach applies to discrete credit migration data and employs two key assumptions:

1. Future rating transitions are independent of past ratings (Markov assumption);
2. The transition probabilities are solely a function of the distance between dates and are independent of the calendar dates (time-homogeneity assumption).

In this approach, we first calculate an empirical TPM for one year. Then, a later-year TPM is obtained by raising the one-year TPM to a power. For example, assume the one-year TPM as M; then a t-year TPM is given by M^t. In practice, we would like these projected TPMs to be as close as possible to the empirical TPMs, $\bar{M}_t$, obtained from empirical data. However, a one-year ETPM may not have a good prediction to the multi-year ETPM and it itself may not satisfy the desired properties of structural stability and the constraints on the default probabilities.

It is clear that we prefer a one-year TPM that satisfies the structural stability constraints and default constraints, and when pushed through time minimizes the deviation from the multi-year ETPMs. Assume that the last row and the last column of an $n \times n$ TPM are associated with the state of default, while other rows and columns correspond to normal credit ratings. We can represent the calculation of a smoothing one-year TPM M in the form of a constrained error minimization problem as follows:

$$\min_{M} \sum_{t=1}^{T} w_t f(M^t, \bar{M}_t, w_{i,j}) \tag{9.1}$$

subject to

$$0 \leq m_{t,ij} \leq 1, \sum_{j=1}^{N} m_{t,ij} = 1, \quad \text{for } t = 1, \tag{9.2}$$

$$m_{t,nn} = 1, \quad \text{and } m_{t,ni} = 0 \text{if } i < n, \quad \text{for } t = 1, \tag{9.3}$$

and

$$m_{t,ij} \leq m_{t,ik}, \quad \text{if } \; j < k \leq i \text{ or } i < k < j < n, \tag{9.4}$$

$$m_{t,in} \leq m_{t,jn}, \quad \text{if } \; i < j, \tag{9.5}$$

for $\forall t \in \{1,2,\cdots,T\}$ and $\forall i,j \in \{1,2,\cdots,n\}$, where T is the number of years of interest. In (9.1), f is an error function that measures the dependency between M^t and $\bar{M}_t$, and w_t is the weight for the dependency at the t-th year. The i-th row and j-th column elements, $m_{t,ij}$ and $\bar{m}_{t,ij}$, of M^t and $\bar{M}_t$ represent the predicted and empirical transition probabilities from the i-th to the j-th credit rating over a t-year period, respectively. w_{ij} is the weight for the dependency between $m_{t,ij}$ and $\bar{m}_{t,ij}$. By varying the weights, the optimization can be customized to emphasize defaults, transitions, or specific rating categories, according to business needs.

The equality constraints in (9.2) ensure that each row of a transition probability matrix sums up to 1. Equation (9.3) implies that the rating of default is an absorbing state, i.e., once an obligor reaches the default state, it is assumed to remain there indefinitely (in practice, an obligor that emerges from default is treated as an entirely new obligor). Thus, the problem in (9.1) is actually to optimize over $n(n-1)$ variables. Note that the constraints (9.2) and (9.3) are automatically satisfied for $t > 1$, as long as they hold for $t = 1$. Eq. (9.4) formulates the structural constraints such that the elements of a TPM are monotonically decreasing on either side of the diagonal. This acts as a mechanism to smooth the probability surface, as shown in Figure 9.2. Equation (9.5) states that a higher rating has a lower probability of default than a lower rating. Different from constraints (9.2) and (9.3), constraints (9.4) and (9.5) do not necessarily hold for multi-year TPM (i.e., $t > 1$) even when they are satisfied by the one-year TPM. However, our experiments (including all the results reported in this chapter) on a wide variety of test data show that these constraints are implicitly satisfied by the optimized TPM for later years once they hold for $t = 1$. For simplicity, we explicitly impose these constraints for

$t = 1$ in this chapter, while assuming that the error minimization procedure implicitly enforces them for later-year TPMs.

We consider three types of error functions, the square, the exponential and the probit [165], to measure the discrepancy between the predicted probability and the empirical data. The corresponding minimization problems are given as follows:

$$\min_M \sum_{t=1}^{T} w_t \sum_{i=1}^{N} \sum_{j=1}^{N} w_{ij} \left(m_{t,ij} - \bar{m}_{t,ij}\right)^2, \tag{9.6}$$

$$\min_M \sum_{t=1}^{T} w_t \sum_{i=1}^{N} \sum_{j=1}^{N} w_{ij} \frac{e^{\left|m_{t,ij} - \bar{m}_{t,ij}\right|} - 1}{e^{\bar{m}_{t,ij}}}, \tag{9.7}$$

and

$$\min_M \sum_{t=1}^{T} w_t \sum_{i=1}^{n} \sum_{j=1}^{n} w_{ij} \left| \Phi^{-1}\left(\frac{m_{t,ij} + \varepsilon}{2}\right) - \Phi^{-1}\left(\frac{\bar{m}_{t,ij} + \varepsilon}{2}\right) \right|, \tag{9.8}$$

where Φ^{-1} is the probit inverse function, and $\varepsilon = 0.0005$ is a small positive number to maintain the robustness of problem when the argument of Φ^{-1} is close to 0. Each of these functions has some advantages and disadvantages in measuring the discrepancy: e.g., the square and exponential functions are relatively easy to computer, while the probit function taker longer; the square function treats the discrepancy on an absolute scale, while the exponential and probit functions work on a relative scale.

9.3 Utilization of Domain Knowledge

As explained in the previous section, the TPM problems are highly nonlinear optimization problems with large problem dimension and a large number of equality and inequality constraints. For example, we considered a problem with $n = 22$ credit ratings. The problem dimension is 462 and the number of constraints is 924, ignoring constraints in (9.5). Due to the complexity of the underlying problem, a penalty function based approach for constraint enforcement in EA was deemed impractical.

To overcome the difficulty, we propose a representation scheme that makes monotonicity constraint satisfaction, as shown in Eqs. (9.4) and (9.5), inherent to the optimization problem. In addition, the problem dimension is reduced to $(n-1)^2$ by transforming the equality constraints of (9.2) into an inherent property of the representation methods. The optimization problem thus obtained has only boundary constraints so that an EA can easily generate only feasible solutions at all times.

We represent a solution as a vector of length $(n-1)^2$, where each element $\in [0, 1]$. For notational convenience, let us first reshape this vector to an $(n-1) \times (n-1)$ matrix X. The meaning of each element x_{ij} of X is described as follows: the last-column element of X represents a ratio between the adjacent last-column elements of an $n \times n$ one-year TPM M (to satisfy the monotonic constraints in (9.5)), and each element in the first $n-2$ columns of X represents a ratio between adjacent row elements of M (to satisfy the monotonic constraints in (9.4)). A TPM matrix is then constructed in the five steps:

Table 9.2 A representation scheme to automatically enforce constraints: *cumprod*(x) returns the cumulative product of an input vector x

Scheme: Construct a solution $M = \{m_{ij}\}$, $i, j = 1, 2, \ldots, n$, from a matrix $X = \{x_{ij}\}$, $i, j = 1, 2, \ldots, n-1$.

$m_{nn} = 1$; $m_{ni} = 0$ for $i = 1, 2, \ldots, n-1$.
Set $(y_{n-1}, y_{n-2} \ldots y_1) = cumprod(x_{n-1,n-1}, x_{n-2,n-1} \ldots x_{1,n-1})$
$m_{in} = y_i$, $i = 1, 2, \ldots, n-1$
For $i = 1 : n$
    $m_{ii} = 1$;
    For $j = i+1 : n-1$
        $m_{ij} = m_{i(j-1)} \times x_{i,j-1}$
    End
    For $j = i-1 : -1 : 1$
        $m_{ij} = m_{i(j+1)} \times x_{i,j}$
    End
    $m_{ij} = m_{ij} / \sum_{j=1}^{n-1} m_{ij} \times (1 - m_{in})$, $j = 1, 2, \ldots, n-1$
End

1. Set the elements in the last row of M according to 9.3;
2. Set the diagonal terms of M to be 1;
3. Calculate the probability terms in the last column of M according the ratio terms in the last column of X;
4. Calculate the off-diagonal terms (except the last column element) in each row of M according the corresponding ratio terms in each row of X (except the last column element);
5. Normalize the first $n-1$ elements in each row of M so that the row adds up to 1

The detailed operations are presented in Table 9.2, where *cumprod* calculates the cumulative product of an input vector. This representation scheme automatically satisfies all the monotonicity constraints in (9.4) and (9.5). Furthermore, there is a one-to-one mapping between a solution vector and the corresponding TPM matrix represented.

9.4 Simulation

In this section, we consider the TPM optimization problem based on the annual ratings data from Standard and Poor's for over 6,000 obligors over the time period from 1989 to 2005 [167]. This data contains 21 normal credit ratings and the rating of the default, and thus the respective TPM matrix to be optimized has $n = 22$ rows and columns. The dimensions D of this problem is $n(n-1)$, i.e., 462. If not mentioned otherwise, the weights defined in the objective function (9.1) are set to be $w_t = 1$ for any t, and $w_{ij} = 0.1$ for $j \neq n$ and $w_{ij} = 5$ for $j = n$. The values of w_{ij} imply a much higher priority of the default probabilities than other transition probabilities.

Table 9.3 The best function values obtained by JADE after 20,000 generations

	Square	Exponential	Probit
Best values	0.2366	8.3016	10.3999

For convenience, we summarize in Table 9.3 the optimal results obtained by JADE after 20,000 generations when different error functions are considered as the objective function. JADE works best among the several algorithms studied below and usually approaches the optimum very fast; thus the values in Table 9.3 may serve as reference points to estimate the performance of different algorithms after a moderate number of generations.

We compare JADE (without archive) with the PSO [31], the classic DE/rand/1/bin [1], as well as jDE [22]. The parameters of JADE are set to be $p = 0.05$ and $c = 0.1$, as in Chapter 4.4. For fair comparison, we follow the parameter settings of jDE as recommended in the original papers, i.e., $\tau_1 = 0.1$ and $\tau_2 = 0.1$. Both DE and PSO's performance are significantly affected by the control parameter settings. In the case of DE, we test nine different parameter settings with F and $CR \in \{0.2, 0.5, 0.9\}$. Experimental results showed that $F = 0.5$ and $CR = 0.9$ generally leads to the best performance. This is consistent with the recommendations in [1], [9], [10], [22]. In the case of PSO, it is not an easy task to systematically study the effect of its three parameters w, c_1, and c_2. Thus, we compare different parameter settings in the literature [31], [169], [170], [171], [172]. Experimental results showed that PSO works best with $w = 0.6$, $c_1 = c_2 = 1.7$ in [31] for the TPM optimization problem. These parameter values will be used in the simulation.

In all evolutionary algorithms, the population size NP is set to be $5D$. The results reported below are obtained based on 30 independent runs of an evolutionary algorithm.

9.4.1 Performance Comparison

Table 9.4 summarizes the mean and standard deviation of the best values obtained by each evolutionary algorithm after 500 or 2000 generations. For a clearer

Table 9.4 A performance comparison of different evolutionary algorithms

Error Function	Gen	JADE	jDE	DE/rand/1/bin	PSO
Square	500	**2.42e-1 (5.96e-4)**	4.85e-1 (1.13e-2)	6.57e-1 (1.49e-2)	3.60e-1 (4.22e-2)
	2000	**2.37e-1 (2.19e-5)**	2.59e-1 (2.85e-3)	5.16e-1 (1.23e-2)	3.37e-1 (4.03e-2)
Exponential	500	**8.72e+0 (2.74e-2)**	1.30e+1 (2.10e-1)	1.55e+1 (2.30e-1)	1.11e+1 (5.36e-1)
	2000	**8.43e+0 (1.82e-2)**	9.21e+0 (9.53e-2)	1.37e+1 (2.14e-1)	1.05e+1 (5.73e-1)
Probit	500	**1.09e+1 (5.55e-2)**	1.62e+1 (3.29e-1)	1.89e+1 (2.44e-1)	1.39e+1 (8.09e-1)
	2000	**1.06e+1 (3.57e-2)**	1.18e+1 (2.44e-1)	1.69e+1 (2.97e-1)	1.31e+1 (8.04e-1)

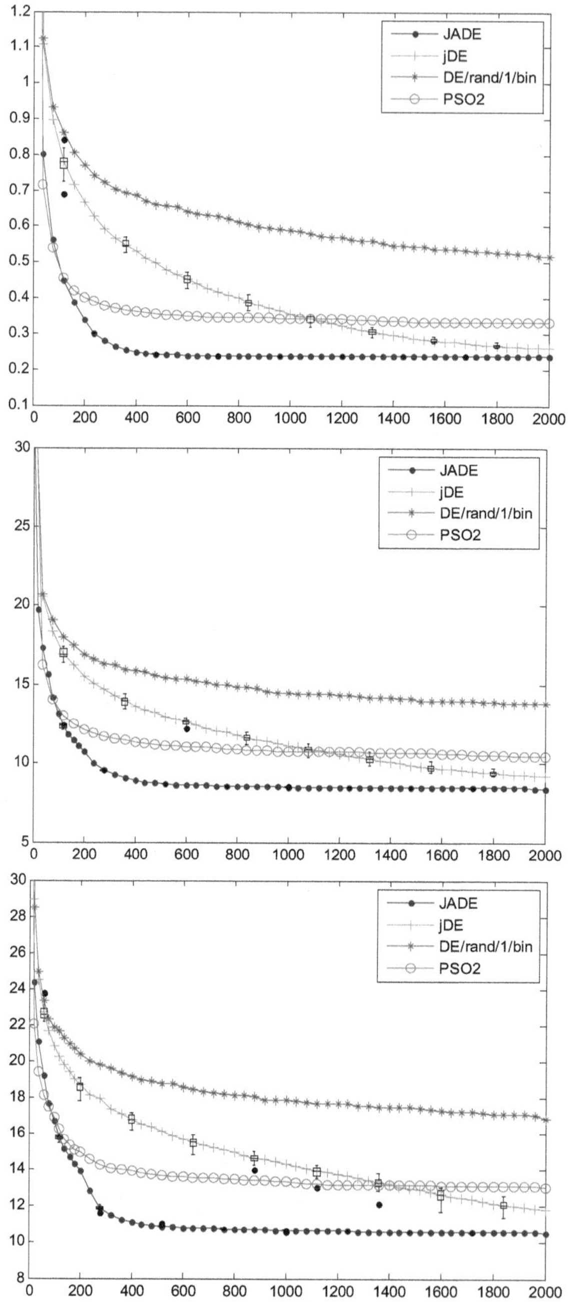

Fig. 9.1 Convergence graphs of different evolutionary algorithms. The x-axis is the number of generations, and the y-axis is the median of function values over 30 independents runs

comparison among evolutionary algorithms, we further plot their convergence graphs in Figure 9.1 In this figure, each curve shows the median of the objective function values obtained after each generation of an evolutionary algorithm in 30 independent runs. It is useful to see how fast an algorithm minimizes the objective function value and whether it converges to a better solution than other algorithms. Other than the curve of median values, box-and-whisker diagrams are plotted at certain generations for the first and second best algorithms. This is helpful to illustrate the spread of results in 30 independent runs and indicate the robustness of the algorithms.

It is clear from Table 9.4 and Figure 9.1 that JADE works best in terms of both convergence rate and robustness for this set of TPM optimization problems. JADE generally obtains near-optimal values in 500 generations, compared to the values achieved after 20,000 generations as summarized in Table 9.3. As a comparison, jDE usually approaches the optimal value after 2,000 generations and the classic DE/rand/1/bin converges even slower. PSO has difficulty solving the TPM problems due to premature convergence, although its convergence rate is fastest during the early generations.

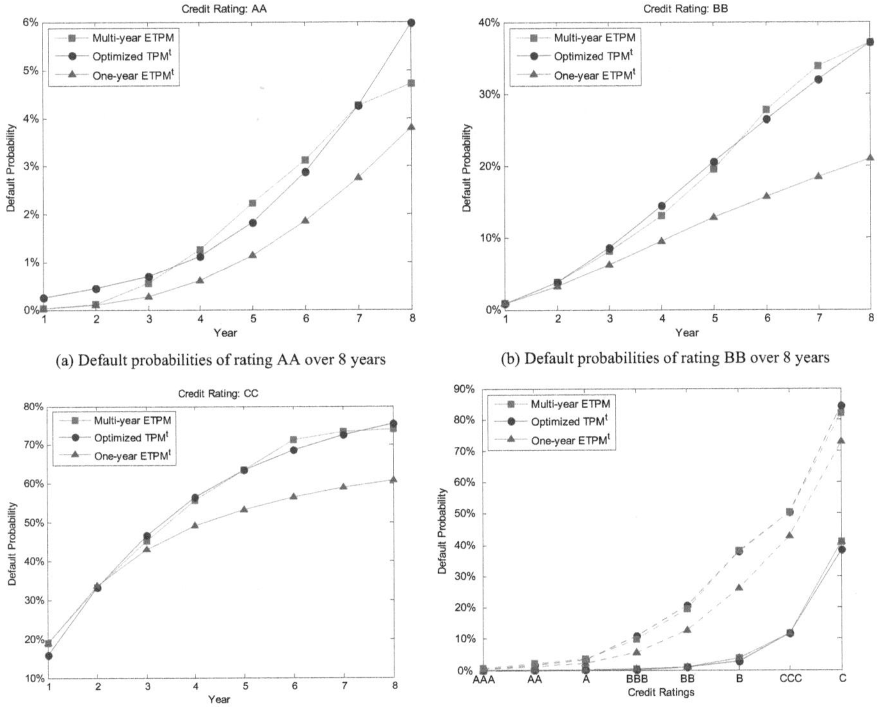

(a) Default probabilities of rating AA over 8 years

(b) Default probabilities of rating BB over 8 years

Fig. 9.2 A comparison of default probabilities obtained from the multi-year ETPM and from the powers of the optimized TPMs obtained using exponential error function. Plots (a) - (c) show the default probabilities of a certain credit rating (e.g., AA, BB, or CC) over 8 years. Plot (d) shows the one-year or three-year default probabilities of multiple credit ratings

In the rest of this section, we focus on the optimization with an exponential error function. Among different error functions discussed in [165], the exponential function is relatively easy to compute and treats the discrepancy on a relative scale.

9.4.2 Comparison of Optimized TPM with Empirical Data

In the following, we compare the results obtained by JADE and the cohort approach. To be specific, we consider the one-year empirical TPM calculated by a cohort method and the optimized one-year TPM obtained by the proposed mythology. The t-year TPM calculated by raising one-year (empirical or optimized) TPM to the t-th power are compared to the target t-year empirical TPMs.

As an example, we first focus on the default structure over an 8-year period. The choice of 8 years as the horizon over which the optimization is performed is arbitrary and does not limit the generality of our method. In practical applications, users are free to set the time horizon that is most relevant to the application where the TPMs are required. As is shown in Figure 9.2, the default probabilities calculated from the cohort TPM are much lower than the empirical default probabilities. The default structure of the optimized TPM is very close to the empirical curves in nearly every period. This shows that the forecast bias is greatly reduced by using the optimized TPM as compared to the cohort TPM.

It is also interesting to observe the shape of the default structure. For practical business reasons, the line should be convex for higher quality rate classes (say AA), gradually become straight for intermediate ratings (say BB), and eventually become concave for lower quality ratings (say CC). It is clear from plots (a) – (c) of Figure 9.2 that the optimized TPM satisfies these desired properties very well, although the multi-year empirical TPMs do not consistently exhibit these properties.

The above results are obtained based on the priority of optimizing the default probabilities by setting a much larger weight for them. As expected, the obtained

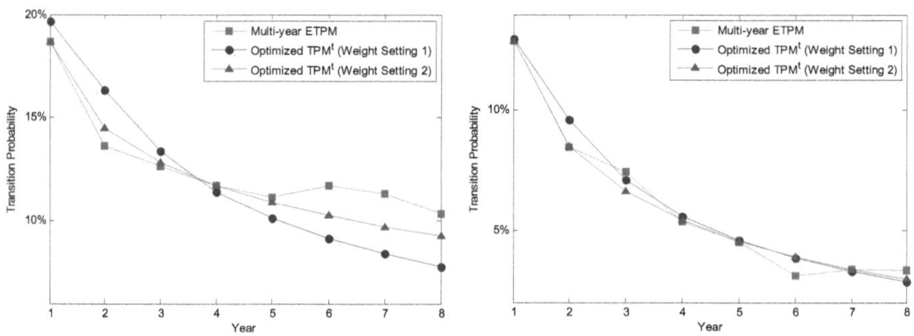

Fig. 9.3 A comparison of transition probabilities from one credit rating to its adjacent rating when different weights are used in the objective function. Left plot: the transition from BBB+ to BBB. Right plot: the transition from B+ to B. Weight setting 1: a small weight of 0.1 is applied to these transition probabilities. Weight setting 2: a relatively large weight of 5 is applied to these transition probabilities

TPM may not have a good prediction for other transitions, e.g., from rating BBB+ to rating BBB. This is the case as shown in Figure 9.3. The emphasis can be changed, however, by applying more weight on specific rating categories according to business needs. For example, we may set $w_{ij} = 5$ if $j = n$ or $|i - j| \leq 1$, and $w_{ij} = 0.1$ otherwise. This weight setting is useful for emphasizing both the default probability and the transition probabilities to adjacent ratings. The simulation results obtained using this setting (weight setting 2) are compared to those obtained using the previous setting (weight setting 1). It is clear from Figure 9.3 that the transition probabilities predicted by optimized TPM have a better match with empirical data, when their corresponding weights get larger.

9.5 Summary

In this chapter, we have considered the problem of smoothing a one-year transition probability matrix by minimizing the discrepancy between predicted later-year TPMs and empirical data over the time horizon of interest. This minimization problem is complex not only in its non-convex non-separable properties but also in the large number of involved variables and constraints (desired properties). By leveraging domain knowledge, we represent the solution in a simple yet efficient form and thus transform the equality and inequality constraints to boundary constraints. Then, the parameter adaptive differential evolution algorithm is adopted to calculate the optimal solution. Simulation results show that the JADE algorithm performs significantly better than other evolutionary algorithms in terms of both the convergence rate of the optimization and the quality of the final solutions obtained. In addition, the default term structure of the optimized TPM is much closer to the empirical curves than the one obtained by the traditional cohort method.

Chapter 10
Conclusions and Future Work

This final chapter presents a summary of conclusions and directions for future research.

10.1 Summary

The fundamental contributions of this research work include the development and analysis of parameter adaptive differential evolution and its application to a variety of real-world problems. The major contributions of the research work are summarized below.

First, a theoretical analysis has been conducted to investigate the evolutionary stochastic properties of a differential evolution algorithm's convergence behavior for a spherical function model. A Gaussian approximate model of differential evolution is introduced to facilitate mathematical derivation of the stochastic properties of mutation and selection and help the understanding of the effect of mutation factor on the evolutionary dynamics. Theoretical results, as verified by experimental simulations, highlight the necessity to adaptively evolve control parameters as evolutionary search proceeds.

Second, a new parameter adaptive differential evolution algorithm, JADE, is introduced to automatically adapt control parameters responding to a range of function characteristics at different stages of evolutionary search. To improve convergence performance, JADE incorporates a greedy mutation strategy that utilizes the direction information provided by both high-quality solutions in the current generation and inferior solutions previously explored in the evolutionary search. Extensive experiments have been conducted to show their mutual benefits for a good balance between the convergence speed and reliability. In addition, the parameter adaptation avoids the need for users' prior knowledge of the interaction between the control parameters and the convergence behavior, thus making the algorithm easily usable.

Third, extensions of the basic JADE algorithm are considered from several perspectives. Some optimization problems have a demanding limitation on the number of original function evaluations that are expensive in terms of time, cost and/or other

J. Zhang and A.C. Sanderson: Adaptive Differential Evolution, ALO 1, pp. 147–150.
springerlink.com

limited resources. JADE has been extended to address these problems by incorporating computationally inexpensive surrogate models and adaptively controlling the level of incorporation according to the model accuracy. Radial basis function networks are used to create these models for the sake of a good balance between computational complexity and model accuracy. Experimental results have shown the efficiency of adaptive incorporation of surrogate models in avoiding potential false or premature convergence while significantly reducing the number of original expensive function evaluations.

Fourth, many real-world applications involve multi-objective optimization, because they are linked to competing or conflicting outcomes such as profit, risk and/or many other performance related criteria. JADE is extended to multi-objective optimization to search for a set of optimal tradeoff solutions. Simulation results show that JADE achieves fast convergence rate by utilizing the direction information provided by previously explored inferior solutions, even when most or all solutions in the current population become non-dominated and thus the high-quality solutions among them do not necessarily indicate the direction of the optimum.

Fifth, the performance of JADE has been evaluated for a variety of benchmark functions of different characteristics such as nonlinearity, non-convexity, multi-modality and high-dimensionality. The performance of JADE is compared with both classic and state-of-the-art algorithms such as the classic DE, other adaptive DE algorithms (SaDE, jDE, SaNSDE), the canonical PSO, Adaptive LEP and Best Levy for single-objective optimization, and MODE, SPEA, GDE3 and NSGA-II for multi-objective optimization. Comprehensive experimental studies have shown that JADE achieves better convergence performance than other algorithms and has better scalability for high-dimensional functions. JADE usually leads to the fastest convergence rate while maintaining the probability of successful convergence at a level very close to the most robust competitive algorithm.

The proposed adaptive differential evolution algorithm JADE has been applied in different real-world applications where domain knowledge can be utilized to improve optimization performance. First, it is applied to a winner determination problem in combinatorial auctions, which can be generalized to resource allocation problems in sensor management, supply chain management, etc. Second, JADE solves a data-smoothing problem in credit decision making by searching for the transition probability matrix that satisfies different desired properties and optimally matches the empirical data of credit rating transition. Furthermore, JADE is applied to conduct automatic flight planning in air traffic control systems. It is used to assign an appropriate route to each flight to minimize both system level congestion and total flight distance. In these applications, JADE has produced better results than other state-of-the-art evolutionary algorithms and heuristic approaches in terms of the convergence rate and the quality of the final solutions.

10.2 Future Work

There are several interesting directions for further research and development based on the work in this monograph.

10.2.1 Coevolutionary Algorithms

The proposed parameter adaptive differential evolution algorithm JADE has shown promising performance for a variety of high-dimensional benchmark functions without explicitly addressing challenges introduced by high dimensionality. In the literature, high-dimensional problems are usually explicitly addressed by "divide and conquer" techniques such as cooperative coevolution (CC) [104], [102], [103], [173], [174]. In general, cooperative coevolution evolves a set of sub-populations, each of which optimizes over a subset of the decision variables by a certain underlying evolutionary algorithm such as DE. To be specific, the underlying evolutionary algorithm is responsible for optimizing over each subset of decision variables, while the interaction or incorporation among different subsets of variables is controlled by a coevolutionary mechanism [175].

Parameter adaptation and cooperative coevolution have quite different emphases on improving optimization performance and their operations are relatively independent of each other. Therefore, it is expected that both classic and adaptive DE algorithms can serve as the underlying evolutionary algorithm for cooperative coevolution. While most existing work adopts classic algorithms as the underlying scheme, a few studies [103] have been conducted to show that the performance of cooperative coevolution can be remarkably improved by taking an adaptive DE algorithm, such as SaNSDE [25], as the basis. Since JADE has shown the best convergence performance scalability for a variety of problems, it is expected that the incorporation of the cooperative coevolution and the parameter adaptation scheme proposed in JADE can produce even better results in solving high-dimensional optimization problems.

10.2.2 Constrained Optimization

In this monograph, we have mainly analyzed optimization problems with no constraints or simple boundary constraints. In the case of the highly constrained TPM optimization problem, we have equivalently transformed the equality and inequality constraints to boundary constraints and thus solve the problem very efficiently by the proposed parameter adaptive algorithm. However, it is usually a great challenge for optimization algorithms (both evolutionary algorithms and conventional nonlinear programming techniques) to solve problems with a large number of linear or nonlinear constraints that cannot be equivalently simplified or loosely relaxed. Highly constrained problems can make optimization extremely difficult because they can create irregular forbidden regions on the search space that restrict the free movement of vectors. The feasible search space may be a non-convex region in the case of nonlinear equality constraints or non-convex inequality constraints. Furthermore, it is not uncommon that constraints may divide the search space into disjoint islands and the proportion of feasible solutions in the whole search space may be negligible. All these challenges will require extensive studies of differential evolution for such complex constrained optimization problems.

10.2.3 Optimization in a Noisy Environment

Optimization in a noisy environment occurs in diverse domains of engineering applications, especially in the case of experimental optimizations where the evaluation of the function involves measuring some physical or chemical data. The function values obtained are affected by noise such as stochastic measurement errors and/or quantization errors. Although the underlying function may be smooth, the function landscape available to the optimization algorithm may show many undesirable properties such as discontinuity and multimodality. In the literature, researchers have mainly been concerned with two types of noise: (i) noise in the fitness function and (ii) noise in the design variables. In the first case, the noise is additive (and in most cases unbiased) to the fitness function value and the research focus is to reduce the influence of the noise. In the second case, the noise is in the design variables and the main motivation is to find an optimal solution that is insensitive to noise (i.e., search for robust solutions).

The optimization of noisy functions has recently drawn more attention and been developed as an independent research topic [90]. Resampling and thresholding are two well-known methods to overcome the noisy fitness evaluation. Resampling suggests evaluation of the same candidate solution N times and approximation of the true fitness value by averaging. The thresholding method is applied in the selection operation, where a parent can only be replaced by its offspring if the fitness value of the offspring is larger than that of the parent by a threshold value τ. The main problem for these two methods is to find an optimal static value of N or τ or update their values adaptively according to the feedback in the optimization process. In this monograph, we have investigated the performance of JADE in a noisy environment without explicitly addressing the noise problem. It looks promising to incorporate such techniques as resampling and thresholding into the proposed adaptive DE algorithm, since the parameter adaptation and the noise reduction have different emphasis on improving the optimization performance and their operations are relatively independent. This will be an interesting topic for future research.

References

1. Storn, R.M., Price, K.V.: Differential evolution - a simple and efficient adaptive scheme for global optimization over continuous spaces, Technical Report TR-95-012, ICSI (March 1995)
2. Price, K.V., Storn, R.M., Lampinen, J.A.: Differential Evolution: A Practical Approach to Global Optimization, 1st edn. Springer, Heidelberg (2005)
3. Rechenberg, I.: Evolutionsstrategie. Frommann-Holzboog, Stuttgart (1973)
4. Schwefel, H.-P.: Evolution and Optimum Seeking. Wiley, New York (1994)
5. Holland, J.H.: Adaptation in Natural and Artificial Systems: An Introductory Analysis with Applications to Biology, Control, and Artificial Intelligence, 3rd edn. MIT Press, Cambridge (1992); First edition: 1975 The University of Michigan
6. Goldberg, D.E.: Genetic Algorithms in Search, Optimization, and Machine Learning. Addison-Wesley Professional, Reading (1989)
7. Fogel, L.J., Owens, A.J., Walsh, M.J.: Artificial Intelligence through Simulated Evolution. John Wiley, Chichester (1966)
8. Nelder, J.A., Mead, R.: A simplex method for function minimization. Computer Journal 7, 308–313 (1965)
9. Mezura-Montes, E., Velázquez-Reyes, J., Coello Coello, C.A.: A comparative study of differential evolution variants for global optimization. In: Genetic and Evolutionary Computation Conference, Seattle, Washington, July 2006, pp. 485–492 (2006)
10. Vesterstroem, J., Thomsen, R.: A comparative study of differential evolution, particle swarm optimization, and evolutionary algorithms on numerical benchmark problems. In: IEEE Congress on Evolutionary Computation, June 2004, pp. 1980–1987 (2004)
11. Back, T.: Evolutionary Algorithms in Theory and Practice: Evolution Strategies, Evolutionary Programming, Genetic Algorithms. Oxford University Press, Oxford (1996)
12. Beyer, H.G.: Theory of Evolution Strategies, 1st edn. Springer, Heidelberg (2001)
13. Zaharie, D.: Control of population diversity and adaptation in differential evolution algorithms. In: Matousek, R., Osmera, P. (eds.) 9th International Conference on Soft Computing, MENDEL (2003)

14. Xue, F., Sanderson, A.C., Graves, R.J.: Modeling and convergence analysis of a continuous multi-objective differential evolution algorithm. In: Proc. of IEEE Congress on Evolutionary Computation, September 2005, vol. 1, pp. 228–235 (2005)
15. Xue, F.: Multi-Objective Differential Evolution: Theory and Applications. Ph.D. dissertation, Rensselaer Polytechnic Institute, Ney York, USA (2004)
16. Abbass, H.A.: The self–adaptive Pareto differential evolution algorithm. In: Proc. of IEEE Congress on Evolutionary Computation, May 2002, vol. 1, pp. 831–836 (2002)
17. Teo, J.: Exploring dynamic self-adaptive populations in differential evolution. Soft Computation: Fusion Found Methodol Appl. 10(8), 673–686 (2006)
18. Liu, J., Lampinen, J.: A fuzzy adaptive differential evolution algorithm. Soft Computation: Fusion Found Methodol Appl. 9(6), 448–462 (2005)
19. Liu, J., Lampinen, J.: Adaptive parameter control of differential evolution. In: Proc. of MENDEL 2002, 8th International Mendel Conference on Soft Computing, Brno, Czech Republic, June 2002, pp. 19–26 (2002)
20. Xue, F., Sanderson, A.C., Bonissone, P.P., Graves, R.J.: Fuzzy logic controlled multi-objective differential evolution. In: Proc. of IEEE International Conference on Fuzzy Systems, June 2005, pp. 720–725 (2005)
21. Qin, A.K., Suganthan, P.N.: Self-adaptive differential evolution algorithm for numerical optimization. In: Proc. of IEEE Congress on Evolutionary Computation, September 2005, vol. 2, pp. 1785–1791 (2005)
22. Brest, J., Greiner, S., Boskovic, B., Mernik, M., Zumer, V.: Self-adapting control parameters in differential evolution: a comparative study on numerical benchmark problems. IEEE Trans. on Evolutionary Computation 10(6), 646–657 (2006)
23. Brest, J., Boskovic, B., Greiner, S., Zumer, V., Maucec, M.S.: Performance comparison of self-adaptive and adaptive differential evolution algorithms. Soft Computation: Fusion Found Methodol Appl. 11(7) (2007)
24. Brest, J., Zumer, V., Maucec, M.S.: Self-Adaptive differential evolution algorithm in constrained real-parameter optimization. In: Proc. of IEEE Congress on Evolutionary Computation, July 2006, pp. 215–222 (2006)
25. Yang, Z., Tang, K., Yao, X.: Self-adaptive differential evolution with neighborhood search. In: Proc. of IEEE Congress on Evolutionary Computation, Hongkong, China (2008)
26. Zhang, J., Sanderson, A.C.: An approximate Gaussian model of differential evolution with spherical fitness functions. In: Proc. of IEEE Congress on Evolutionary Computation (CEC 2007), Singapore, September 2007, pp. 2220–2228 (2007)
27. Zhang, J., Sanderson, A.C.: JADE: self-adaptive differential evolution with fast and reliable convergence performance. In: Proc. of IEEE Congress on Evolutionary Computation (CEC 2007), Singapore, September 2007, pp. 2251–2258 (2007)
28. Zhang, J., Sanderson, A.C.: JADE: adaptive differential evolution with optional external archive. IEEE Trans. on Evolutionary Computation (2009) (accepted for publication)
29. Zhang, J., Sanderson, A.C.: DE-AEC: a differential evolution algorithm based on adaptive evolution control. In: Proc. of IEEE Congress on Evolutionary Computation (CEC 2007), Singapore, September 2007, pp. 3824–3830 (2007)

30. Zhang, J., Sanderson, A.C.: Self-adaptive multi-objective differential evolution with direction information provided by archived inferior solutions. In: Zurada, J.M., Yen, G.G., Wang, J. (eds.) Computational Intelligence: Research Frontiers. LNCS, vol. 5050. Springer, Heidelberg (2008)
31. Trelea, I.C.: The particle swarm optimization algorithm: Convergence analysis and parameter selection. Information Processing Letters 85(6), 317–325 (2003)
32. Yao, X., Liu, Y., Lin, G.: Evolutionary programming made faster. IEEE Trans. on Evolutionary Computation 3(2), 82–102 (1999)
33. Lee, C.Y., Yao, X.: Evolutionary programming using mutations based on the Lévy probability distribution. IEEE Trans. on Evolutionary Computation 8(1), 1–13 (2004)
34. Kukkonen, S., Lampinen, J.: GDE3: The third evolution step of generalized differential evolution. In: Proc. of IEEE Congress on Evolutionary Computation, Edinburgh, Scotland, September 2005, pp. 443–450 (2005)
35. Deb, K., Agrawal, S., Pratap, A., Meyarivan, T.: A fast and elitist multi-objective genetic algorithm: NSGA-II. IEEE Trans. Evolutionary Computation 6(2), 182–197 (2002)
36. Zhang, J., Sanderson, A.C.: On the scalability of adaptive differential evolution algorithms for high-dimensional problems. IEEE Trans. Evolutionary Computation (submitted, 2009)
37. Zhang, J., Avasarala, V., Sanderson, A.C., Mullen, T.: Differential evolution for discrete optimization: an experimental study on combinatorial auction problems. In: Zurada, J.M., Yen, G.G., Wang, J. (eds.) Computational Intelligence: Research Frontiers. LNCS, vol. 5050. Springer, Heidelberg (2008)
38. Zhang, J., Avasarala, V., Subbu, R.: Evolutionary optimization of transition probability matrices for credit decision-making. European Journal of Operational Research (2009) (accepted for publication)
39. Darwin, C.: The Origin of Species. John Murray (1859)
40. Bertsekas, D.P.: Nonlinear Programming, 2nd edn. Athena Scientific (1999)
41. Fogel, D.B.: An introduction to simulated optimization. IEEE Trans. on Neural Networks 5(1), 3–14 (1994)
42. Fogel, D.B.: Evolving Artificial Intelligence, PhD thesis, University of California, San Diego, CA (1992)
43. De Jong, K.A.: An Analysis of the Behavior of a Class of Genetic Adaptive Systems, Ph.D. thesis, University of Michigan, Ann Arbor, Michigan (1975)
44. Back, T., Schwefel, H.-P.: Evolutionary computation: An overview. In: Proceedings of the IEEE International Conference on Evolutionary Computation, Nagoya, Japan (1996)
45. Michalewicz, Z.: Genetic Algorithms + Data Structure = Evolution Programs, 3rd edn. Springer, New York (1996)
46. Koza, J.R.: Genetic Programming: on the Programming of Computers by Means of Natural Selection. MIT Press, Cambridge (1992)
47. Angeline, P.J.: Adaptive and self-adaptive evolutionary computations. In: Computational Intelligence: A Dynamic Systems Perspective, pp. 152–163 (1995)
48. Eiben, A.E., Hinterding, R., Michalewicz, Z.: Parameter control in evolutionary algorithms. IEEE Trans. on Evolutionary Computation 3(2), 124–141 (1999)
49. Eiben, A.E., Smith, J.E.: Introduction to Evolutionary Computing. Springer, New York (2003)

50. Subbu, R., Bonissone, P.: A retrospective view of fuzzy control of evolutionary algorithm behavior. In: Proc. of IEEE International Conference on Fuzzy Systems, St. Louis, MO, USA (May 2003)
51. Coello Coello, C.A., Lamont, G.B.: Applications of Multi-Objective Evolutionary Algorithms. World Scientific, Singapore (2004)
52. Deb, K.: Multi-Objective Optimization using Evolutionary Algorithms. John Wiley & Sons, Chichester (2001)
53. Hwang, C.L., Masud, A.S.: Multiple Objective Decision Making — Methods and Applications. Springer, Berlin (1979)
54. Van Veldhuizen, D.A., Lamont, G.B.: Multiobjective evolutionary algorithms: analyzing the state-of-the-art. Evolutionary Computation 8(2), 125–147 (2000)
55. Coello Coello, C.A., Pulido, G.T., Montes, E.M.: Current and future research trends in evolutionary multiobjective optimization. In: Information Processing with Evolutionary Algorithms: From Industrial Applications to Academic Speculations, pp. 213–231. Springer, Heidelberg (2005)
56. Adra, S.F., Griffin, I., Fleming, P.J.: A comparative study of progressive preference articulation techniques for multiobjective optimisation. In: Obayashi, S., Deb, K., Poloni, C., Hiroyasu, T., Murata, T. (eds.) EMO 2007. LNCS, vol. 4403, pp. 908–921. Springer, Heidelberg (2007)
57. Coello Coello, C.A., Lamont, G.B., Van Veldhuizen, D.A.: Evolutionary Algorithms for Solving Multi-Objective Problems, 2nd edn. Springer, Heidelberg (2007)
58. Knowles, J.D., Corne, D.W.: Approximating the nondominated front using the Pareto archived evolution strategy. Evolutionary Computation 8(2), 149–172 (2000)
59. Zitzler, E., Thiele, L.: Multiobjective evolutionary algorithms: A comparative case study and the strength Pareto approach. IEEE Trans. on Evolutionary Computation 3(4), 257–271 (1999)
60. Zitzler, E., Laumanns, M., Thiele, L.: SPEA2: Improving the strength Pareto evolutionary algorithm. In: Computer Engineering and Networks Laboratory (TIK), Swiss Federal Inst. Technology (ETH), Zurich, Switzerland, 103 (2001)
61. Corne, D.W., Knowles, J.D., Oates, M.J.: The Pareto envelopebased selection algorithm for multiobjective optimization. In: Deb, K., Rudolph, G., Lutton, E., Merelo, J.J., Schoenauer, M., Schwefel, H.-P., Yao, X. (eds.) PPSN 2000. LNCS, vol. 1917, pp. 839–848. Springer, Heidelberg (2000)
62. Wolpert, D.H., MacReady, W.G.: No free lunch theorems for optimization. IEEE Trans. on Evolutionary Computation 1(1), 67–82 (1997)
63. Igel, C., Toussaint, M.: A no-free-lunch theorem for non-uniform distributions of target functions. Journal of Mathematical Modelling and Algorithms 3, 313–322 (2004)
64. Ho, Y.-C., Pepyne, D.L.: Simple explanation of the no free lunch theorem of optimization. Cybernetics and Systems Analysis 38(2), 292–298 (2002)
65. Bonissone, P.P., Subbu, R., Eklund, N., Kiehl, T.R.: Evolutionary algorithms + domain knowledge = real-world evolutionary computation. IEEE Trans. on Evolutionary Computation 10(3), 256–280 (2006)
66. Arnold, D.V., Beyer, H.-G.: Investigation of the (μ, λ)-ES in the presence of noise. In: Proc. of the IEEE Congress on Evolutionary Computation, pp. 332–339 (2001)

67. Arnold, D.V., Beyer, H.-G.: On the benefits of populations for noisy optimization. Evolutionary Computation 11(2), 111–127 (2003)
68. Beyer, H.G.: On the performance of (1, λ)-evolution strategies for the ridge function class. IEEE Trans. on Evolutionary Computation 5(3), 218–235 (2001)
69. Arnold, D.V., Beyer, H.-G.: Evolution strategies with cumulative step length adaptation on the noisy parabolic ridge, Technical Report CS-2006-02, Dalhousie University (2006)
70. Beyer, H.G.: Towards a theory of 'evolution strategies': progress rates and quality gain for (1,+λ)-strategies on (nearly) arbitrary fitness functions. In: Davidor, Y., Männer, R., Schwefel, H.-P. (eds.) PPSN 1994. LNCS, vol. 866, pp. 58–67. Springer, Heidelberg (1994)
71. Hansen, N., Ostermeier, A.: Completely derandomized self-adaptation in evolution strategies. Evolutionary Computation 9(2), 159–195 (2001)
72. Gamperle, R., Muller, S.D., Koumoutsakos, P.: A parameter study for differential evolution. In: Advances in Intelligent Systems, Fuzzy Systems, Evolutionary Computation, pp. 293–298. WSEAS Press (2002)
73. Davies, R.B.: Algorithm AS155: the distribution of a linear combination of (chi-squared) random variables. Applied Statistics 29(3), 323–333 (1980)
74. Farebrother, W.: Remark AS R53: A remark on algorithms AS 106, AS 153 and AS 155: The distribution of a linear combination of $\chi 2$ random variables. Applied Statistics 33(3), 366–369 (1984)
75. Patel, J.K., Read, C.B.: Handbook of the Normal Distribution, 2nd edn. Statistics, a Series of Textbooks and Monographs. CRC, Boca Raton (1996)
76. Gnedenko, B., Kolmogorov, A.: Limit Distributions for sums of Independent Random Variables. Addson-Wesley, Reading (1968)
77. Mendes, R., Rocha, I., Ferreira, E.C., Rocha, M.: A comparison of algorithms for the optimization of fermentation processes. In: Proc. of IEEE Congress on Evolutionary Computation, July 2006, pp. 2018–2025 (2006)
78. Eberhart, R.C., Shi, Y., Kennedy, J.: Swarm Intelligence, 1st edn. Morgan Kaufmann, San Francisco (2001)
79. Tvrdik, J., Krivy, I., Misik, L.: Evolutionary algorithm with competing heuristics. In: Proceedings of MENDEL 2001, 7th International Conference on Soft Computing, Brno, Czech, June 2001, pp. 58–64 (2001)
80. Tvrdík, J., Misik, L., Krivy, I.: Competing heuristics in evolutionary algorithms. Intelligent Technologies–Theory and Applications, 159–165 (2002)
81. Huang, V.L., Qin, A.K., Suganthan, P.N.: Self-adaptive differential evolution algorithm for constrained real-parameter optimization. In: Proc. of IEEE Congress on Evolutionary Computation, July 2006, pp. 17–24 (2006)
82. Babu, B.V., Jehan, M.M.L.: Differential evolution for multi-objective optimization. In: Proc. of IEEE Congress on Evolutionary Computation, December 2003, vol. 4, pp. 2696–2703 (2003)
83. Pahner, U., Hameyer, K.: Adaptive coupling of differential evolution and multiquadricsapproximation for the tuning of the optimization process. IEEE Trans. on Magnetics 36(4), 1047–1051 (2000)
84. Mezura-Montes, E., Velazquez-Reyes, J., Coello Coello, C.A.: Modified differential evolution for constrained optimization. In: Proc. of IEEE Congress on Evolutionary Computation, July 2006, pp. 25–32 (2006)
85. Yao, X., Liu, Y., Liang, K.-H., Lin, G.: Fast evolutionary algorithms. In: Rozenberg, G., Back, T., Eiben, A. (eds.) Advances in Evolutionary Computing: Theory and Applications, pp. 45–94. Springer, New York (2003)

86. Shang, Y.-W., Qiu, Y.-H.: A note on the extended rosenbrock function. Evolutionary Computation 14(1), 119–126 (2006)
87. Dixon, L.C.W., Szegö, G.: The global optimization problem: An introduction. In: Dixon, L.C.W., Szegö, G. (eds.) Toward Global Optimization 2, pp. 1–15. North-Holland, Amsterdam (1978)
88. The MathWorks, Inc., `http://www.mathwork.com`
89. Yang, Z., He, J., Yao, X.: Making a difference to differential evolution. In: Michalewicz, Z., Siarry, P. (eds.) Advances in Metaheuristics for Hard Optimization, pp. 397–414. Springer, Heidelberg (2007)
90. Jin, Y., Branke, J.: Evolutionary optimization in uncertain environments - A survey. IEEE Trans. on Evolutionary Computation 9(3), 303–317 (2005)
91. Krink, T., Filipic, B., Fogel, G.B.: Noisy optimization problems - A particular challenge for differential evolution? In: Proc. of the 2004 Congress on Evolutionary Computation, vol. 1, pp. 332–339 (2004)
92. Das, S., Konar, A., Chakraborty, U.K.: Improved differential evolution algorithms for handling noisy optimization problems. In: Proc. of IEEE Congress on Evolutionary Computation, vol. 2, pp. 1691–1698 (2005)
93. Rahnamayan, S., Tizhoosh, H.R., Salama, M.M.A.: Opposition-dased differential evolution for optimization of noisy problems. In: Proc. of IEEE Congress on Evolutionary Computation, July 2006, pp. 1865–1872 (2006)
94. Hansen, N., Niederberger, A.S.P., Guzzella, L., Koumoutsakos, P.: A method for handling uncertainty in evolutionary optimization with an application to feedback control of combustion. IEEE Trans. on Evolutionary Computation (accepted)
95. Craig, J.W.: A new, simple, and exact result for calculating the probability of error for two-dimensional signal constellations. In: Proc. IEEE Military Communications Conf., McLean, VA, October 1991, pp. 571–575 (1991)
96. Bousmalis, K., Pfaffmann, J., Hayes, G.: Improving evolutionary algorithms with scouting: high-dimensional problems. In: Rutkowski, L., Tadeusiewicz, R., Zadeh, L.A., Zurada, J.M. (eds.) ICAISC 2008. LNCS, vol. 5097, pp. 365–375. Springer, Heidelberg (2008)
97. Gao, Y., Wang, Y.-J.: A memetic differential evolutionary algorithm for high dimensional functions' optimization. In: International Conference on Natural Computation, August 2007, vol. 4, pp. 188–192 (2007)
98. Helwig, S., Wanka, R.: Particle swarm optimization in high-dimensional bounded search spaces. In: IEEE Swarm Intelligence Symposium, April 2007, pp. 198–205 (2007)
99. Li, H.-Q., Li, L.: A novel hybrid particle swarm optimization algorithm combined with harmony search for high dimensional optimization problems. In: International Conference on Intelligent Pervasive Computing, October 2007, pp. 94–97 (2007)
100. Zheng, X., Chen, K., Lu, D., Liu, H.: A proposal for a cooperative coevolutionary PSO. In: IEEE International Symposium on Information Technologies and Applications in Education, November 2007, pp. 324–329 (2007)
101. Suganthan, P.N., Hansen, N., Liang, J.J., Deb, K., Chen, Y.P., Auger, A., Tiwari, S.: Problem definitions and evaluation criteria for the CEC 2005 special session on real-parameter optimization, Technical Report, Nanyang Technological University, Singapore (2005)

102. Shi, Y., Teng, H., Li, Z.: Cooperative co-evolutionary differential evolution for function optimization. In: Wang, L., Chen, K., S. Ong, Y. (eds.) ICNC 2005. LNCS, vol. 3611, pp. 1080–1088. Springer, Heidelberg (2005)
103. Yang, Z., Tang, K., Yao, X.: Large scale evolutionary optimization using cooperative coevolution. Information Sciences 178(15), 2985–2999 (2008)
104. Liu, Y., Yao, X., Zhao, Q., Higuchi, T.: Scaling up fast evolutionary porgramming with cooperative coevolution. In: Proc. of IEEE Congress on Evolutionary Computation, pp. 1101–1108 (2001)
105. Jin, Y.: A comprehensive survey of fitness approximation in evolutionary. Soft Computing 9(1), 3–12 (2003)
106. Jones, D.R.: A taxonomy of global optimization methods based on response surface. J. of Global Optim. 21(4), 345–383 (2001)
107. Zhou, Z., Ong, Y.S., Lim, M.H., Lee, B.S.: Memetic algorithm using multi-surrogates for computationally expensive optimization problems. Soft Computing 11(10), 957–971 (2006)
108. Ong, Y.S., Nair, P.B., Keane, A.J.: Evolutionary optimization of computationally expensive problems via surrogate modeling. American Inst. of Aeronautics and Astronautics J. 41, 687–696 (2003)
109. Regis, R.G., Shoemaker, C.A.: Improved strategies for radial basis function methods for global optimization. Journal of Global Optimization 37(1), 113–135 (2007)
110. Regis, R.G., Shoemaker, C.A.: Local function approximation in evolutionary algorithms for the optimization of costly functions. IEEE Trans. on Evolutionary Computation 8(5), 490–505 (2004)
111. Knowles, J.: ParEGO: A hybrid algorithm with on-line landscape approximation for expensive multiobjective optimization problems. IEEE Trans. on Evolutionary Computation 10(1), 50–66 (2006)
112. Jin, Y., Olhofer, M., Sendhoff, B.: On evolutionary optimization with approximate fitness functions. In: Proc. of the Genetic and Evolutionary Computation Conference, pp. 786–792. Morgan Kaufmann, San Francisco (2000)
113. Powell, M.J.D.: The theory of radial basis function approximation in 1990. In: Light, W. (ed.) Advances in Numerical Analysis. Wavelets, Subdivision Algorithms and Radial Basis Functions, vol. 2, pp. 105–210. Oxford Univ. Press, London (1992)
114. Powell, M.J.D.: Recent research at cambridge on radial basis functions. In: Muller, M., Buhmann, M., Mache, D., Felten, M. (eds.) New Developments in Approximation Theory. International Series of Numerical Mathematics, vol. 132, pp. 215–232. Birkhauser Verlag, Basel (1999)
115. Giannakoglou, K.C.: Design of optimal aerodynamic shapes using stochastic optimizationmethods and computational intelligence. Int. Rev. J. Progress Aerosp. Sci. 38(5), 43–76 (2002)
116. Ye, K.Q., Li, W., Sudjianto, A.: Algorithmic construction of orthogonal symmetric latin hypercube designs. J. Statis. Planning Inference 90, 145–159 (2000)
117. Liang, J.J., Runarsson, T.P., Mezura-Montes, E., Clerc, M., Suganthan, P.N., Coello Coello, C.A., Deb, K.: Problem definitions and evaluation criteria for the CEC 2006, Special Session on Constrained Real-Parameter Optimization, Technical Report, Nanyang Technological University, Singapore (2006)
118. Silverman, B.W.: Density Estimation for Statistics and Data Analysis. Chapman and Hall, London (1986)

119. Coello Coello, C.A.: 20 years of evolutionary multi-objective optimization: What has been done and what remains to be done. In: Yen, G.Y., Fogel, D.B. (eds.) Computational Intelligence: Principles and Practice, ch. 4, pp. 73–88. IEEE Computational Intelligence Society, Los Alamitos (2006)
120. Srinivas, N., Deb, K.: Multiobjective optimization using nondominated sorting in genetic algorithms. Evolutionary Computation 2(3), 221–248 (1994)
121. Khare, V., Yao, X., Deb, K.: Performance scaling of multi-objective evolutionary algorithms. In: Fonseca, C.M., Fleming, P.J., Zitzler, E., Deb, K., Thiele, L. (eds.) EMO 2003. LNCS, vol. 2632, pp. 376–390. Springer, Heidelberg (2003)
122. Abbass, H.A., Sarker, R., Newton, C.: PDE: A Pareto-frontier differential evolution approach for multi-objective optimization problems. In: Proc. of IEEE Congress on Evolutionary Computation, pp. 971–978 (2001)
123. Abbass, H., Sarker, R.: The Pareto differential evolution algorithm. International Journal of Artificial Intelligence Tools 11(4), 531–552 (2002)
124. Sarker, R., Abbass, H.: Differential evolution for solving multi-objective optimization problems. Asia-Pacific Journal of Operations Research 2(2), 225–240 (2004)
125. Madavan, N.K.: Multiobjective optimization using a pareto differential evolution approach. In: Proc. of IEEE Congress on Evolutionary Computation, pp. 1145–1150 (2002)
126. Xue, F., Sanderson, A.C., Graves, R.J.: Pareto-based multi-objective differential evolution. In: Proc. of IEEE Congress on Evolutionary Computation, Canberra, Australia, pp. 862–869 (2003)
127. Hernandez-Diaz, A.G., Santana-Quintero, L.V., Coello Coello, C.: A new proposal for multi-objective optimization using differential evolution and rough sets theory. In: Proc. of the 8th Annual Conference on Genetic and Evolutionary Computation, pp. 675–682 (2006)
128. Iorio, A., Li, X.: Incorporating directional information within a differential evolution algorithm for multiobjective optimization. In: Proc. of Genetic and Evolutionary Computation Conference, pp. 691–697 (2006)
129. Robic, T., Filipic, B.: DEMO: Differential evolution for multiobjective optimization. In: Coello Coello, C.A., Hernández Aguirre, A., Zitzler, E. (eds.) EMO 2005. LNCS, vol. 3410, pp. 520–533. Springer, Heidelberg (2005)
130. Tušar, T.: Design of an Algorithm for Multiobjective Optimization with Differential Evolution. Master Thesis, University of Ljubljana, Slovenian (2007)
131. Zielinski, K., Laur, R.: Variants of differential evolution for multi-objective optimization. In: IEEE Symposium on Computational Intelligence in Multicriteria Decision Making, April 2007, pp. 91–98 (2007)
132. Kukkonen, S., Deb, K.: Improved pruning of non-dominated solutions based on crowding distance for bi-objective optimization problems. In: Proc. of IEEE Congress on Evolutionary Computation, pp. 1179–1186 (2006)
133. Kukkonen, S., Deb, K.: A fast and effective method for pruning of non-dominated solutions in many-objective problems. In: Runarsson, T.P., Beyer, H.-G., Burke, E.K., Merelo-Guervós, J.J., Whitley, L.D., Yao, X. (eds.) PPSN 2006. LNCS, vol. 4193, pp. 553–562. Springer, Heidelberg (2006)
134. Huang, V.L., Suganthan, P.N., Baskar, S.: Multiobjective differential evolution with external archive and harmonic distance-based diversity measure, Technical Report, Nanyang Technological University, Singapore (December 2005)

135. Mo, J., Walrand, J.: Fair end-to-end windows-based congestion control. IEEE/ACM Trans. On Networking 8(5), 556–567 (2000)
136. Deb, K., Thiele, L., Laumanns, M., Zitzler, E.: Scalable test problems for evolutionary multiobjective optimization. In: Evolutionary Multiobjective Optimization, pp. 105–145. Springer, London (2005)
137. Huband, S., Hingston, P., Barone, L., While, L.: A review of multi-objective test problems and a scalable test problem toolkit. IEEE Trans. on Evolutionary Computation 10(5), 477–506 (2007)
138. Becerra, R.L., Coello Coello, C.A., Hernández-Díaz, A.G., Caballero, R., Molina, J.: Alternative techniques to solve hard multi-objective optimization Problems. In: Proc. of the 9th Annual Conference on Genetic and Evolutionary Computation, London, England, pp. 754–757 (2007)
139. Rudolph, G., Naujoks, B., Preuss, M.: Capabilities of EMOA to detect and preserve equivalent Pareto subsets. In: Obayashi, S., Deb, K., Poloni, C., Hiroyasu, T., Murata, T. (eds.) EMO 2007. LNCS, vol. 4403, pp. 36–50. Springer, Heidelberg (2007)
140. Okabe, T., Jin, Y., Olhofer, M., Sendhoff, B.: On test functions for evolutionary multiobjective optimization. In: Yao, X., Burke, E.K., Lozano, J.A., Smith, J., Merelo-Guervós, J.J., Bullinaria, J.A., Rowe, J.E., Tiňo, P., Kabán, A., Schwefel, H.-P. (eds.) PPSN 2004. LNCS, vol. 3242, pp. 792–802. Springer, Heidelberg (2004)
141. Knowles, J.D., Thiele, L., Zitzler, E.: A tutorial on the performance assessment of stochastic multi-objective optimizers, Technical Report TIK-Report No. 214, Computer Engineering and Networks Laboratory, ETH Zürich (Feburary 2006)
142. Conover, W.J.: Practical Nonparameteric Statistics, 3rd edn. Hohn Wiley and Sons, New York (1999)
143. Bandyopadhyay, S., Saha, S., Maulik, U., Deb, K.: A simulated annealing based multi-objective optimization algorithm: AMOSA. IEEE Trans. on Evolutionary Computation 12(3), 269–283 (2008)
144. Lampinen, J., Zelinka, I.: Mixed integer-discrete-continuous optimization by differential evolution, Part 1: the optimization method. In: Proc. of MENDEL 1999, 5th International Mendel Conference on Soft Computing, Czech Republic, June 1999, pp. 71–76 (1999)
145. Lampinen, J., Zelinka, I.: Mixed integer-discrete-continuous optimization by differential evolution, part 2: a practical example. In: Proc. of MENDEL 1999, 5th Int. Conf. on Soft Computing, Czech Republic, June 1999, pp. 77–81 (1999)
146. Xue, F., Sanderson, A.C., Graves, R.J.: Multi-objective differential evolution - algorithm, convergence analysis, and applications. In: Proc. of IEEE Congress on Evolutionary Computation, pp. 743–750 (2005)
147. Pan, Q.-K., Tasgetiren, M.F., Liang, Y.-C.: A discrete differential evolution algorithm for the permutation flowshop scheduling problem. In: Proc. of the 9th Annual Conference on Genetic and Evolutionary Computation, pp. 126–133 (2007)
148. Cramton, P., Shoham, Y., Steinberg, R.: Combinatorial Auctions. MIT Press, Cambridge (2006)
149. Avasarala, V., Mullen, T., Hall, D.L., Garga, A.: MASM: market architecture or sensor management in distributed sensor networks. In: SPIE Defense and Security Symposium, Orlando, FL, pp. 5813–5830 (2005)

150. Mullen, T., Avasarala, V., Hall, D.L.: Customer-driven sensor management. IEEE Intelligent Systems 21(2), 41–49 (2006)
151. Walsh, W.E., Wellman, M., Ygge, F.: Combinatorial auctions for supply chain formation. In: Proc. of ACM Conf. on Electronic Commerce, pp. 260–269 (2000)
152. Das, A., Grosu, D.: A combinatorial auction-based protocols for resource allocation in grids. In: 19th IEEE International Parallel and Distributed Processing Symposium (2005)
153. Rothkopf, M., Pekec, A., Harstad, R.: Computationally manageable combinatorial auctions. Management Science 44(8), 1131–1147 (1998)
154. ILOG CPLEX, ILOG Inc., `http://www.ilog.com/products/cplex/`
155. Andersson, A., Mattias, T., Fredrik, Y.: Integer programming for combinatorial auction winner determination. In: Proc. of the Fourth International Conf. Multi-Agent Systems, Boston, MA, pp. 39–46 (2000)
156. Sandholm, T., Subhash, S., Gilpin, A., Levine, D.: CABOB: A fast optimal algorithm for winner determination in combinatorial auctions. In: International Joint Conference on Artificial Intelligence, pp. 1002–1008 (2001)
157. Hoos, H., Boutilier, C.: Solving combinatorial auctions using stochastic local search. In: Proc. of the Seventeenth National Conf. on Artificial Intelligence and Twelfth Conference on Innovative Applications of Artificial Intelligence, pp. 22–29 (2000)
158. Avasarala, V., Polavarapu, H., Mullen, T.: An approximate algorithm for resource allocation using combinatorial auctions. In: Proc. of the IEEE/WIC/ACM international conference on Intelligent Agent Technology, pp. 571–578 (2006)
159. Kennedy, J., Eberhart, R.C.: Particle swarm optimization. In: Proc. of IEEE International Conference on Neural Networks, Piscataway, NJ, pp. 1942–1948 (1995)
160. Kennedy, J., Eberhart, R.C.: A discrete binary version of particle swarm optimization. In: Proc. of the Conf. on Systems, Man, and Cybernetics, Piscataway, NJ, pp. 4104–4109 (1997)
161. Engelbrecht, A.P., Pampara, G.: Binary differential evolution strategies. In: Proc. of the IEEE Congress on Evolutionary Computation, Singapore, September 2007, pp. 1942–1947 (2007)
162. Veeramachaneni, K., Osadciw, L., Kamath, G.: Probabilistically driven particle swarms for discrete multi valued problems: design and analysis. In: IEEE Swarm Intelligence Symposium, Hawaii (April 2007)
163. Crook, I., Tibichte, Z., Subbu, R., Lizzi, J., Zhang, J., Jha, P., Suchkov, A.: User-oriented traffic flow planning for the next generation air transport system. In: 7th AIAA Aviation Technology, Integration and Operations Conference, Northern Ireland (September 2007)
164. Subbu, R., Lizzi, J., Iyer, N., Jha, P.D., Suchkov, A.: MONACO–multi-objective national airspace collaborative optimization. In: IEEE Aerospace Conference, March 2007, pp. 1–14 (2007)
165. Chalermkraivuth, K., Keenan, S., Neagu, R., Ellis, J.A., Black, J.W.: Generating transition probability matrices through an optimization framework. In: Quantitative Methods in Finance Conference, Sydney, Australia (2007)
166. Cantor, R., Packer, F.: Determinants and impacts of sovereign credit ratings. Economic Policy Review 2(2), 37–53 (1996)

167. Vazza, D., Aurora, D., Erturk, E.: Annual 2006 global corporate default study and rating transitions, Standard and Poor's 2006 (2006), http://www2.standardandpoors.com/spf/pdf/fixedincome/AnnualDefaultStudy_2005.pdf
168. Jafry, Y., Schuermann, T.: Measurement and estimation of credit migration matrices. Journal of Banking and Finance 288(11), 2603–2639 (2004)
169. Kennedy, J., Eberhart, R.C., Shi, Y.: Swarm Intelligence, 1st edn. Morgan Kaufmann, San Francisco (2001)
170. Shi, Y., Eberhart, R.: Empirical study of particle swarm optimization. In: Proc. of IEEE Congress on Evolutionary Computation, pp. 1945–1950 (1999)
171. Carlisle, A., Dozier, G.: An off-the-shelf PSO. In: Proc. of the Workshop on Particle Swarm Optimization, pp. 1–6 (2001)
172. Eberhart, R.C., Shi, Y.: Comparing inertia weights and constriction factors in particle swarm optimization. In: Proc. of IEEE Congress on Evolutionary Computation, pp. 84–88 (2000)
173. Yang, Z., Tang, K., Yao, X.: Differential evolution for high-dimensional function optimization. In: Proc. of IEEE Congress on Evolutionary Computation, September 2007, pp. 3523–3530 (2007)
174. Olorunda, O., Engelbrecht, A.P.: Differential evolution in high-dimensional search spaces. In: Proc. of IEEE Congress on Evolutionary Computation, September 2007, pp. 1934–1941 (2007)
175. Subbu, R., Sanderson, A.C.: Network-Based Distributed Planning Using Coevolutionary Algorithms. World Scientific Publishing Company, Singapore (2004)

Index

FSC
www.fsc.org
MIX
Papier | Fördert
gute Waldnutzung
FSC® C083411

Kolibri 360 GmbH
Ferdinand-Jühlke-Straße 7
99095 Erfurt, Deutschland
produktsicherheit@kolibri360.de